21世纪高等学校规划教材 | 计算机科学与技术

Java EE技术实验教程

韩姗姗 王春平 编著

清华大学出版社
北京

内 容 简 介

本书以社会对Java EE技术开发人才的要求为目标,以轻量级Java EE编程技术为重点,共分成了4个部分:Web程序设计知识回顾、轻量级Java EE编程技术、企业级EJB组件编程技术和Java EE综合应用开发,总计15个实验。

其中,第一部分回顾已学的JSP、Servlet、JavaBean、JDBC等技术,共有1个实验;第二部分主要针对Struts2、Hibernate和Spring基本开发方法、框架核心机制和程序设计思想进行练习,共包括9个实验;第三部分主要针对会话Bean、实体Bean和消息驱动Bean开发技术进行练习,共包括3个实验;第四部分设计了2个综合实验,分别要求综合运用SSH框架或EJB框架进行系统设计开发。

本书前三个部分中的每个实验都包括3个层次的小实验:基础实验、提高实验和扩展实验,其知识范围和实现难度递增,可供教师和学生自主选择。

本书为学校计算机类专业的实验教学量身定制,可作为高校Java EE技术开发相关课程的教材,也可作为社会中的Java EE技术培训班的教材。本书读者对象应有Java程序设计以及JSP、Servlet、JDBC等JavaWeb的相关基础开发知识。

本书封面贴有清华大学出版社防伪标签,无标签者不得销售。
版权所有,侵权必究。举报: 010-62782989,beiqinquan@tup.tsinghua.edu.cn。

图书在版编目(CIP)数据

Java EE技术实验教程/韩姗姗,王春平编著. --北京:清华大学出版社,2015(2024.2重印)
21世纪高等学校规划教材·计算机科学与技术
ISBN 978-7-302-40068-4

Ⅰ.①J… Ⅱ.①韩…②王… Ⅲ.①JAVA语言-程序设计-高等学校-教材 Ⅳ.①TP312

中国版本图书馆CIP数据核字(2015)第089869号

责任编辑:郑寅堃 王冰飞
封面设计:傅瑞学
责任校对:李建庄
责任印制:宋 林

出版发行:清华大学出版社
网　　址:https://www.tup.com.cn,https://www.wqxuetang.com
地　　址:北京清华大学学研大厦A座　　　　邮　编:100084
社 总 机:010-83470000　　　　　　　　　　邮　购:010-62786544
投稿与读者服务:010-62776969,c-service@tup.tsinghua.edu.cn
质量反馈:010-62772015,zhiliang@tup.tsinghua.edu.cn
课件下载:https://www.tup.com.cn,010-62795954

印 装 者:三河市龙大印装有限公司
经　　销:全国新华书店
开　　本:185mm×260mm　　印 张:13.5　　字　数:339千字
版　　次:2015年7月第1版　　　　　　　　印　次:2024年2月第10次印刷
印　　数:3661~4160
定　　价:35.00元

产品编号:064922-02

出版说明

随着我国改革开放的进一步深化,高等教育也得到了快速发展,各地高校紧密结合地方经济建设发展需要,科学运用市场调节机制,加大了使用信息科学等现代科学技术提升、改造传统学科专业的投入力度,通过教育改革合理调整和配置了教育资源,优化了传统学科专业,积极为地方经济建设输送人才,为我国经济社会的快速、健康和可持续发展以及高等教育自身的改革发展做出了巨大贡献。但是,高等教育质量还需要进一步提高以适应经济社会发展的需要,不少高校的专业设置和结构不尽合理,教师队伍整体素质亟待提高,人才培养模式、教学内容和方法需要进一步转变,学生的实践能力和创新精神亟待加强。

教育部一直十分重视高等教育质量工作。2007年1月,教育部下发了《关于实施高等学校本科教学质量与教学改革工程的意见》,计划实施"高等学校本科教学质量与教学改革工程"(简称"质量工程"),通过专业结构调整、课程教材建设、实践教学改革、教学团队建设等多项内容,进一步深化高等学校教学改革,提高人才培养的能力和水平,更好地满足经济社会发展对高素质人才的需要。在贯彻和落实教育部"质量工程"的过程中,各地高校发挥师资力量强、办学经验丰富、教学资源充裕等优势,对其特色专业及特色课程(群)加以规划、整理和总结,更新教学内容、改革课程体系,建设了一大批内容新、体系新、方法新、手段新的特色课程。在此基础上,经教育部相关教学指导委员会专家的指导和建议,清华大学出版社在多个领域精选各高校的特色课程,分别规划出版系列教材,以配合"质量工程"的实施,满足各高校教学质量和教学改革的需要。

为了深入贯彻落实教育部《关于加强高等学校本科教学工作,提高教学质量的若干意见》精神,紧密配合教育部已经启动的"高等学校教学质量与教学改革工程精品课程建设工作",在有关专家、教授的倡议和有关部门的大力支持下,我们组织并成立了"清华大学出版社教材编审委员会"(以下简称"编委会"),旨在配合教育部制定精品课程教材的出版规划,讨论并实施精品课程教材的编写与出版工作。"编委会"成员皆来自全国各类高等学校教学与科研第一线的骨干教师,其中许多教师为各校相关院、系主管教学的院长或系主任。

按照教育部的要求,"编委会"一致认为,精品课程的建设工作从开始就要坚持高标准、严要求,处于一个比较高的起点上。精品课程教材应该能够反映各高校教学改革与课程建设的需要,要有特色风格、有创新性(新体系、新内容、新手段、新思路,教材的内容体系有较高的科学创新、技术创新和理念创新的含量)、先进性(对原有的学科体系有实质性的改革和发展,顺应并符合21世纪教学发展的规律,代表并引领课程发展的趋势和方向)、示范性(教材所体现的课程体系具有较广泛的辐射性和示范性)和一定的前瞻性。教材由个人申报或各校推荐(通过所在高校的"编委会"成员推荐),经"编委会"认真评审,最后由清华大学出版社审定出版。

目前,针对计算机类和电子信息类相关专业成立了两个"编委会",即"清华大学出版社计算机教材编审委员会"和"清华大学出版社电子信息教材编审委员会"。推出的特色精品教材包括:

(1) 21世纪高等学校规划教材·计算机应用——高等学校各类专业,特别是非计算机专业的计算机应用类教材。

(2) 21世纪高等学校规划教材·计算机科学与技术——高等学校计算机相关专业的教材。

(3) 21世纪高等学校规划教材·电子信息——高等学校电子信息相关专业的教材。

(4) 21世纪高等学校规划教材·软件工程——高等学校软件工程相关专业的教材。

(5) 21世纪高等学校规划教材·信息管理与信息系统。

(6) 21世纪高等学校规划教材·财经管理与应用。

(7) 21世纪高等学校规划教材·电子商务。

(8) 21世纪高等学校规划教材·物联网。

清华大学出版社经过三十多年的努力,在教材尤其是计算机和电子信息类专业教材出版方面树立了权威品牌,为我国的高等教育事业做出了重要贡献。清华版教材形成了技术准确、内容严谨的独特风格,这种风格将延续并反映在特色精品教材的建设中。

清华大学出版社教材编审委员会

联系人:魏江江

E-mail:weijj@tup.tsinghua.edu.cn

前 言

Java EE 技术是目前流行的企业级应用开发体系架构,包含软件开发的重要技术标准。Java EE 技术综合了 Java EE 的体系架构、开发模式、程序设计、数据库、网络通信等内容,学习 Java EE 技术的最终目的是将这些理论知识融会贯通来解决实际问题。本书作为理论联系实际的落脚点,旨在将 Java EE 技术包含的主要内容通过实验的形式展现出来,在帮助教师开展实验指导工作的同时,也使得学生能更好地参与实验,并通过实验环节提高动手能力,加深对理论知识的理解,获得分析探索、交流讨论、团队协作、解决问题等可迁徙技能。

本书根据当前用人单位的实际需要,选择轻量级 Java EE 编程技术为重点,分成 4 个部分:Web 程序设计知识回顾、轻量级 Java EE 编程技术、企业级 EJB 组件编程技术、Java EE 综合应用开发,共 15 个实验,如表 0-1 所示。每个实验都包括 3 个层次的小实验:基础实验、提高实验和扩展实验,其知识范围和实现难度递增。

表 0-1 实验体系

第一部分		Web 程序设计知识回顾
	实验一	Servlet 与 JSP 技术——第一个用户登录模块
第二部分		轻量级 Java EE 编程技术
Struts2	实验二	Struts2 基础应用——基于 Struts2 框架的用户登录模块
	实验三	Struts2 的控制器组件 Action——登录用户的功能扩展
	实验四	Struts2 的工作流程——登录用户的高级功能
Hibernate	实验五	Hibernate 基础应用——基于 Hibernate 框架的用户登录模块
	实验六	Hibernate 的体系结构——登录用户信息的增、删、改、查
	实验七	Hibernate 关联关系映射——登录用户的地址管理
Spring	实验八	SSH 整合(Spring4+Struts2+Hibernate4)——基于 SSH 的用户注册模块
	实验九	Spring 的核心机制:控制反转(IoC)——登录用户的购物车
	实验十	Spring 的面向切面编程(AOP)——用户登录模块的增强处理
第三部分		企业级 EJB 组件编程技术
	实验十一	会话 Bean——用会话 Bean 实现用户登录及购物车应用
	实验十二	实体 Bean——用实体 Bean 实现用户信息的持久化
	实验十三	消息驱动 Bean——登录用户支付消息的分发应用
第四部分		Java EE 综合应用开发
	实验十四	综合应用——基于 SSH 的网上书城
	实验十五	综合应用——基于 EJB 的网上书城

本书以社会对 Java EE 技术开发人才的要求为目标,结合新技术的发展,以双线索组织

实验内容。本书的主要特点是：

(1) 实验内容紧密联系社会实际需求。

本书选择目前应用范围最广的，也是社会实际需求面最大的轻量级 Java EE 编程技术——SSH(即 Struts2＋Spring4＋Hibernate4)为重点，以企业级 EJB 组件编程技术为辅形成实验指导教材的主要内容，尽量避免教材内容的大而全和实验技术方案相对落后的缺点。

(2) 以双线索组织实验内容。

本书以双线索组织实验内容。明线是根据项目的构建过程和层次结构进行实验内容组织。教程中的实验都围绕同一个项目模块(用户登录模块)作为基本实验内容进行切入，使得该模块的功能得到不断的丰富和完善。这样的安排保证了实验内容之间的连贯性，也使得学生能够在一个难度和规模适中的模块里，通过逐步添加新的功能，完成新旧知识的联系。暗线是根据学生的能力发展进行实验内容组织。教程中的实验包括内容和难易不同的 3 个层次，分别针对基础开发能力、综合应用能力和可迁移能力的培养。通过实验不但希望提高学生的工程实践能力，而且希望学生的分析探索、交流讨论、团队协作、问题解决等可迁徙技能得到发展。

(3) 实验内容分层可选。

教程中的实验包括基本实验、提高实验和扩展实验 3 个层次。这 3 个层次在内容上相互关联，在难度上层层递进，在能力要求上逐级发展。基础实验难度较低，用于培养学生解决问题的信心和兴趣；提高实验和扩展实验有挑战性，用于激励学生开展进一步的探索和创新。学生可以根据自身的能力水平和知识结构选择不同层次和难度的实验，这既有利于学生开展自主学习，也有利于教师"因材施教"。

(4) 理论与实践相结合。

在每一个实验前，都将介绍实验的目的、基本知识和原理、主要步骤和目标要求，使得本教程成为一个自包含的系统，能够使得学生根据本教程的说明完成实验内容。

本书为学校计算机类专业的实验教学量身定制，可作为高校 Java EE 技术开发相关课程的教材，也可作为社会中的 Java EE 技术培训班的教材。本书读者对象应有 Java 程序设计以及 JSP、Servlet、JDBC 等 JavaWeb 的相关基础开发知识。

本书的第一、三部分由王春平编写，第二、四部分由韩姗姗编写。受作者水平所限，书中的错误和不妥之处在所难免，敬请读者批评指正。

作　者

2015 年 5 月

目 录

第一部分 Web 程序设计知识回顾

实验一　Servlet 与 JSP 技术——第一个用户登录模块 …… 3
　　基础实验——Servlet 与 JSP 基础开发 …… 3
　　提高实验——Servlet 与 JSP 集成的 MVC 方案 …… 9
　　扩展实验——JDBC 与 DAO 设计模式 …… 12

第二部分 轻量级 Java EE 编程技术

实验二　Struts2 基础应用——基于 Struts2 框架的用户登录模块 …… 19
　　基础实验——Struts2 框架搭建 …… 19
　　提高实验——Struts2 标签 …… 23
　　扩展实验——Struts2 的国际化 …… 25

实验三　Struts2 的控制器组件 Action——登录用户的功能扩展 …… 28
　　基础实验——Action 的自定义方法 …… 28
　　提高实验——ActionSupport 与输入校验 …… 31
　　扩展实验——Action 类与 Servlet API …… 37

实验四　Struts2 的工作流程——登录用户的高级功能 …… 42
　　基础实验——拦截器与过滤器 …… 42
　　提高实验——值栈与 OGNL …… 47
　　扩展实验——Struts2 的异常处理 …… 50

实验五　Hibernate 基础应用——基于 Hibernate 框架的用户登录模块 …… 54
　　基础实验——Hibernate 框架搭建 …… 54
　　提高实验——持久化对象与 Hibernate 映射文件 …… 60
　　扩展实验——粒度设计 …… 65

实验六　Hibernate 的体系结构——登录用户信息的增、删、改、查 …… 72
　　基础实验——Hibernate 常用 API …… 72
　　提高实验——HQL 语言 …… 79

扩展实验——深入 Hibernate 配置文件 ………………………………………… 83

实验七　Hibernate 关联关系映射——登录用户的地址管理 ……………………… 86

基础实验——一对多/多对一关联 ……………………………………………… 86
提高实验——多对多关联 ………………………………………………………… 92
扩展实验——一对一关联 ………………………………………………………… 95

实验八　SSH 整合（Spring4+Struts2+Hibernate4）——基于 SSH 的用户注册模块 …… 100

基础实验——Spring 框架搭建 ………………………………………………… 100
提高实验——Spring 与 Hibernate 的整合 …………………………………… 104
扩展实验——Spring、Struts 与 Hibernate 的整合 …………………………… 108

实验九　Spring 的核心机制：控制反转（IoC）——登录用户的购物车 ………… 111

基础实验——Spring 容器中的依赖注入 ……………………………………… 111
提高实验——Spring 容器中的 Bean …………………………………………… 115
扩展实验——深入 Spring 容器 ………………………………………………… 120

实验十　Spring 的面向切面编程（AOP）——用户登录模块的增强处理 ……… 125

基础实验——使用@AspectJ 实现 AOP ………………………………………… 125
提高实验——使用 Spring AOP 实现事务管理 ………………………………… 129
扩展实验——Spring AOP 的核心工作原理：代理和代理工厂 ……………… 133

第三部分　企业级 EJB 组件编程技术

实验十一　会话 Bean——用会话 Bean 实现用户登录及购物车应用 ………… 139

基础实验——无状态会话 Bean 的调用 ………………………………………… 139
提高实验——有状态会话 Bean 的调用 ………………………………………… 148
扩展实验——控制会话 Bean 的生命周期 ……………………………………… 152

实验十二　实体 Bean——用实体 Bean 实现用户信息的持久化 ………………… 157

基础实验——实体 Bean 的开发 ………………………………………………… 157
提高实验——使用 JPQL 语言 …………………………………………………… 165
扩展实验——实体关系映射操作 ………………………………………………… 169

实验十三　消息驱动 Bean——登录用户支付消息的分发应用 ………………… 177

基础实验——处理点对点消息 …………………………………………………… 177
提高实验——处理发布/订阅消息 ………………………………………………… 181
扩展实验——支付消息的同步和异步订阅 ……………………………………… 184

第四部分 Java EE 综合应用开发

实验十四 综合应用——基于 SSH 的网上书城 ·············· 193

实验十五 综合应用——基于 EJB 的网上书城 ·············· 203

第一部分　Web程序设计知识回顾

- 实验一　Servlet与JSP技术
　　　　——第一个用户登录模块

实验一

Servlet 与 JSP 技术
——第一个用户登录模块

基础实验——Servlet 与 JSP 基础开发

一、实验目的

1. 掌握 HttpServlet 的概念、相关 API 以及开发步骤。
2. 掌握 JSP 技术的基本语法。
3. 掌握 JSP 各隐含变量的使用方法。
4. 掌握使用 Servlet 和 JSP 集成开发简单用户登录功能的方法。

二、基本知识与原理

1. Servlet 是用于实现 Web 应用程序设计的 Java 技术解决方案,旨在扩展 Web 服务器的功能。它是由 Servlet 容器(例如 Tomcat)创建并管理的。

2. JSP(Java Server Pages)页面是包含 Java 代码和 HTML 标签的 Web 页面。它是由 JSP 标签和 HTML 标签混合而成的 Web 页面,主要用于用户交互。

三、实验内容及步骤

1. 下载并解压安装 Eclipse 的 Java EE 集成开发环境,如图 1-1 所示。

图 1-1 下载并解压安装 Eclipse 的 Java EE 集成开发环境

2. 在 Eclipse 中新建动态 Web 工程(Dynamic Web Project) javaweb-prj1,如图 1-2 所示。
3. 为工程 javaweb-prj1 添加 Servlet 开发的外部库文件 servlet-api.jar。操作步骤如下。
(1) 右击工程 javaweb-prj1 图标,在弹出的菜单中执行 Build Path | Configure Build

Path 命令，如图 1-3 所示。

图 1-2　创建动态 Web 工程

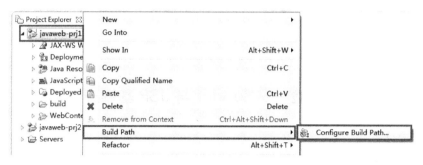

图 1-3　为工程添加外部库文件的菜单操作

（2）在弹出的窗口中选择 Java Build Path|Libraries 选项卡，单击 Add External JARs 按钮，接着在弹出的窗口中切换到 Tomcat 安装目录（例如 C:\Program Files（x86）\Apache Software Foundation\Tomcat 6.0)下的 lib 目录，然后选中并打开加载 servlet-api.jar 库文件，最后单击 OK 按钮，如图 1-4 所示。

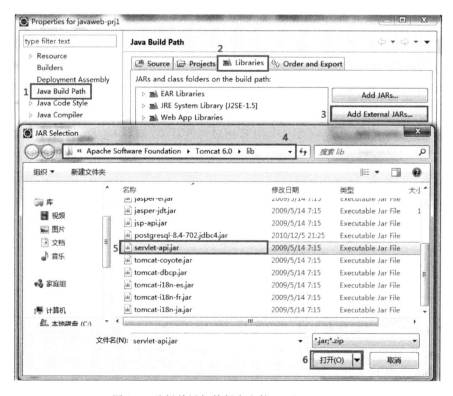

图 1-4　选择并添加外部库文件 servlet-api.jar

4. 在javaweb-prj1中，右击WebContent目录，新建用户登录页面login.jsp，如图1-5所示。

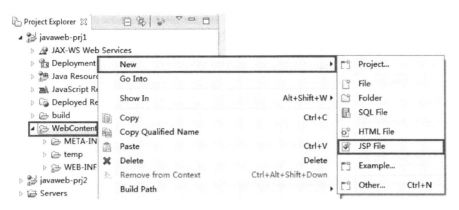

图1-5 创建JSP页面

具体代码如下所示（注意表单form属性action的值，以及用户名与密码输入框中属性name的值）。

```
<%@ page language = "java" contentType = "text/html; charset = GBK"
    pageEncoding = "GBK" %>
<!DOCTYPE html PUBLIC " - //W3C//DTD HTML 4.01 Transitional//EN" "http://www.w3.org/TR/html4/loose.dtd">
<html>
<head>
<meta http - equiv = "Content - Type" content = "text/html; charset = GBK">
<title>用户登录页面</title>
</head>
<body>
<form action = "login" method = "post">
        请输入用户名: < input name = "username" type = "text"><br>
        请输入密码: < input name = "password" type = "password">
                < input type = "submit" value = "登录">
</form>
</body>
</html>
```

5. 在javaweb-prj1中，右击Java Resources|src目录，新建一个名称为cn.edu.zjut的包。然后在该包下创建一个Servlet：LoginController.java（如图1-6所示），用于接收login.jsp页面提交的用户名和密码。如果用户名和密码均为zjut，则输出"登录成功，欢迎您！"，否则输出"用户名或密码错误！"。

具体代码如下所示。

```
package cn.edu.zjut;
import java.io. * ;
import javax.servlet. * ;
import javax.servlet.http. * ;
public class LoginController extends HttpServlet {
    protected void doPost(HttpServletRequest request, HttpServletResponse response) throws
```

```
ServletException, IOException {
        response.setContentType("text/html;charset = utf - 8");
        PrintWriter out = response.getWriter();
        String username = request.getParameter("username");
        String password = request.getParameter("password");
        if("zjut".equals(username) && "zjut".equals(password)){
                out.println("登录成功,欢迎您!");
        }else{
                out.println("用户名或密码错误!");
        }
    }
}
```

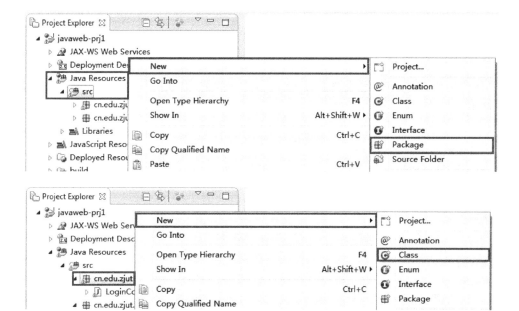

图 1-6　创建 Package 包及 Java 类

6. 在 javaweb-prj1 的 WebContent | WEB-INF 目录下，新建 web.xml 文件，为 LoginController 配置 url-pattern 映射。具体代码如下所示。

```
<?xml version = "1.0" encoding = "UTF - 8"?>
< web - app xmlns:xsi = "http://www.w3.org/2001/XMLSchema - instance"
        xmlns = "http://java.sun.com/xml/ns/javaee"

xmlns:web = "http://java.sun.com/xml/ns/javaee/web - app_2_5.xsd"
        xsi:schemaLocation = "http://java.sun.com/xml/ns/javaee
                http://java.sun.com/xml/ns/javaee/web - app_2_5.xsd"
            id = "WebApp_ID" version = "2.5">
    < display - name > javaweb - prj1 </display - name >
    < servlet >
        < servlet - name > LoginController </servlet - name >
        < display - name > LoginController </display - name >
        < description ></description >
```

```xml
        <servlet-class>cn.edu.zjut.LoginController</servlet-class>
    </servlet>
    <servlet-mapping>
        <servlet-name>LoginController</servlet-name>
        <url-pattern>/login</url-pattern>
    </servlet-mapping>
</web-app>
```

7. 将 javaweb-prj1 部署在 Tomcat 服务器上,具体操作步骤如下。

(1) 下载并安装好 Tomcat(本例为版本 6.0,实际应用中可选择 6.0 以上版本)。

(2) 打开 Eclipse,执行 Window|Preferences 命令,如图 1-7 所示。

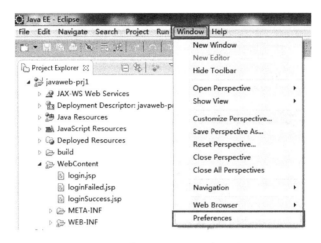

图 1-7　打开 Eclipse 的首选项

(3) 在弹出的对话框的左侧栏中选择 Server|Runtime Environments,然后单击右侧栏的 Add 按钮,如图 1-8 所示。

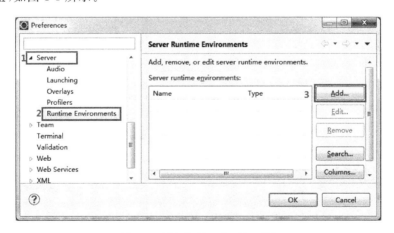

图 1-8　添加新的运行时间环境

(4) 在弹出的对话框中,选择列表框中的 Apache Tomcat v6.0 选项,并单击 Next 按钮,如图 1-9 所示。此时进入 Tomcat 的服务器配置对话框。单击 Browse 按钮,接着在弹出的目录选择对话框中选择 Tomcat 安装目录(例如 C:\Program Files (x86)\Apache

Software Foundation\Tomcat 6.0）。单击"确定"按钮，完成 Tomcat 服务器的选择。然后，依次单击 Finish 按钮和 OK 按钮，保存服务器配置方案，如图 1-10 所示。

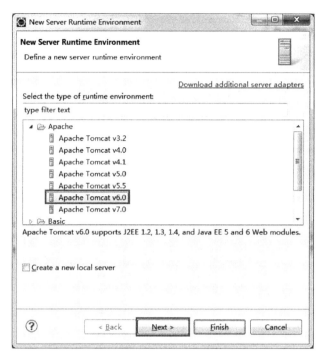

图 1-9　选定 Tomcat 服务器版本

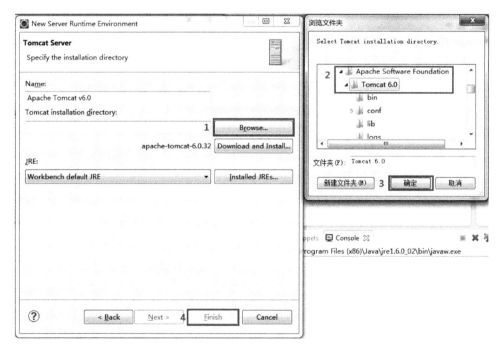

图 1-10　选择实际已安装的 Tomcat 服务器

(5) 右击 javaweb-prj1 工程,在弹出的菜单中执行 Run As|Run on Server 命令。然后在弹出的对话框中选择刚才配置的 Tomcat v6.0 Server,并单击 Finish 按钮,此时服务器会自动启动并加载运行 javaweb-prj1 工程。同时,服务器启动的相关信息会出现在 Console 控制栏,如图 1-11 所示。

图 1-11 Console 控制栏中显示的 Tomcat 服务器启动信息

(6) 打开任一浏览器,输入网址 http://localhost:8080/javaweb-prj1/login.jsp,即可开始访问 login.jsp 页面,进行相关运行调试。

8. 运行 login.jsp 页面,输入用户名和密码,并记录运行结果。

9. 修改 login.jsp 页面,使用表格对表单域进行对齐排列,运行并观察结果。

10. 修改 login.jsp 页面,使用 JavaScript 对用户名表单域 username 和密码表单域 password 进行校验(校验规则:不能为空且不能超过 6 位),运行并观察结果。

四、实验要求

1. 填写并上交实验报告,报告中应包括如下内容:
(1) 运行结果截图;
(2) 修改后的关键代码及相应的运行结果或报错信息;
(3) 实验收获及总结。
2. 上交程序源代码,主要代码中应加注详细的注释。

提高实验——Servlet 与 JSP 集成的 MVC 方案

一、实验目的

1. 掌握 JavaBean 的编写要点。
2. 掌握 MVC 设计模式的 3 个组成要素:模型(Model)、视图(View)和控制器(Controller)。
3. 能在 Web 应用程序设计中熟练使用 MVC 模式。

二、基本知识与原理

1. MVC 模式将交互式应用分成模型(Model)、视图(View)和控制器(Controller) 3 个部分。

2. 模型(Model)是指从现实世界中抽象出来的对象模型,它是应用逻辑的反映。模型封装了数据和数据的操作,是实际进行数据处理计算的地方。

3. 视图(View)是应用和用户之间的接口,它负责将应用显现给用户,并显示模型的状态。

4. 控制器(Controller)负责视图和模型之间的交互,控制对用户输入的响应内容、响应方式和流程。它主要负责两方面的动作:把用户的请求分发到相应的模型,将模型的改变及时反映到视图上。

5. MVC 设计模式将业务逻辑和显示逻辑进行了良好的分离,使代码更加清晰,可维护性更好。

三、实验内容及步骤

1. 在工程 javaweb-prj1 中新建用户模型 UserBean.java,具体代码如下所示。

```java
package cn.edu.zjut.model;
public class UserBean {
    //属性声明
    private String username;
    private String password;
    //构造方法
    public UserBean(){}
    //get 方法
    public String getUsername(){ return username; }
    public String getPassword(){ return password; }
    //set 方法
    public void setUsername(String username){
        this.username = username;
    }
    public void setPassword(String password){
        this.password = password;
    }
}
```

2. 在 javaweb-prj1 中新建视图 loginSuccess.jsp 和 loginFailed.jsp,分别用于显示登录成功和登录失败后的页面。具体代码如下所示。

```jsp
                    <!-- loginSuccess.jsp 页面源代码 -->
<%@ page language = "java" contentType = "text/html; charset = GBK"
    pageEncoding = "GBK" import = "cn.edu.zjut.model.UserBean" %>
<!DOCTYPE html PUBLIC "-//W3C//DTD HTML 4.01 Transitional//EN" "http://www.w3.org/TR/html4/loose.dtd">
<html>
<head>
<meta http-equiv = "Content-Type" content = "text/html; charset = GBK">
<title>登录成功</title>
</head>
<body>
<% UserBean user = (UserBean)request.getAttribute("USER"); %>
登录成功,欢迎您,<% = user.getUsername() %>!
```

```
</body>
</html>
```

```
                    <!-- loginFailed.jsp 页面源代码 -->
<%@ page language = "java" contentType = "text/html; charset = GBK"
    pageEncoding = "GBK" %>
<!DOCTYPE html PUBLIC " - //W3C//DTD HTML 4.01 Transitional//EN" "http://www.w3.org/TR/html4/loose.dtd">
<html>
<head>
<meta http - equiv = "Content - Type" content = "text/html; charset = GBK">
<title>登录失败</title>
</head>
<body>
登录失败,用户名或密码错误!
</body>
</html>
```

3. 修改 javaweb-prj1 中的控制器 LoginController.java,实现如下功能。

(1) 调用 UserBean 用户模型,记录用户信息。

(2) 如果用户名和密码均为 zjut,则跳转到视图 loginSuccess.jsp,否则跳转到视图 loginFailed.jsp,具体代码如下所示。

```
package cn.edu.zjut;
import java.io.*;
import javax.servlet.*;
import javax.servlet.http.*;
import cn.edu.zjut.model.*;

public class LoginController extends HttpServlet {
    public void doPost(HttpServletRequest request, HttpServletResponse response) throws ServletException, IOException {
        String username = request.getParameter("username");
        String password = request.getParameter("password");
        UserBean user = new UserBean();
        user.setUsername(username);
        user.setPassword(password);
        if(checkUser(user)){
            request.setAttribute("USER", user);
            RequestDispatcher dispatcher = request
                .getRequestDispatcher("/loginSuccess.jsp");
            dispatcher.forward(request, response);
        }else{
response.sendRedirect("/javaweb - prj1/loginFailed.jsp");
        }
    }
    boolean checkUser(UserBean user){
        if("zjut".equals(user.getUsername()) &&
            "zjut".equals(user.getPassword())) {
```

```
            return true;
        }else{
            return false;
        }
    }
}
```

4. 将 javaweb-prj1 工程重新部署在 Tomcat 服务器上。

5. 运行 login.jsp 页面,输入用户名和密码并运行,观察运行结果。如果出现运行错误,请尝试修正。

6. 对 javaweb-prj1 工程做如下修改。

(1) 修改视图 login.jsp,新增一个下拉列表框的表单域,用于选择用户类型(值为"普通用户"和"管理员")。

(2) 修改用户模型 UserBean.java,新增一个用户类型 type 属性。

(3) 修改控制器 LoginController.java,要求用户名和密码均为 zjut,并且在用户类型为"管理员"时才跳转到 loginSuccess.jsp 页面,否则跳转到 loginFailed.jsp 页面。

7. 运行并观察修改后的结果,领会采用视图、模型和控制器分离显示逻辑和业务逻辑的优点。

四、实验要求

1. 填写并上交实验报告,报告中应包括如下内容。
(1) 运行结果截图。
(2) 分析实验过程,查找相关资料,总结 MVC 设计模式的优缺点。
(3) 实验收获及总结。
2. 上交程序源代码,代码中应有相关注释。

扩展实验——JDBC 与 DAO 设计模式

一、实验目的

1. 掌握数据库操作的基本 SQL 语句。
2. 掌握连接数据库的基本步骤。
3. 掌握利用 JDBC 操作数据库的基本 API 和方法。
4. 掌握 DAO 设计模式。
5. 掌握在一个项目中集成 MVC 设计模式和 DAO 设计模式的方法。
6. 进一步理解 MVC 设计模型和实体的对应关系。
7. 理解 DAO 设计模式中业务逻辑和数据访问逻辑相分离的优点。

二、基本知识与原理

1. JDBC(Java Data Base Connectivity)是用于执行 SQL 语句的 Java API,负责为多种关系数据库提供统一访问。它由一组用 Java 语言编写的类和接口组成。JDBC 提供了一种基准,用于构建更高级的工具和接口,以帮助数据库开发人员编写数据库应用程序。

2. DAO(Data Access Object)设计模式将所有对数据源的访问操作抽象封装在一组公共 API 接口中,其中定义了应用程序中将会用到的所有事务方法。当程序开发人员需要和数据源进行交互的时候,可直接使用这组接口,无须操作数据库。

三、实验内容及步骤

1. 下载并安装 PostgreSQL 数据库(版本 8.3)。下载地址:http://www.enterprisedb.com/products-services-training/pgdownload#windows。

2. 创建一个用户名为 dbuser,密码为 dbpassword 的数据库登录角色,然后为该角色创建一个名称为 myDB 的数据库,并在该数据库中创建一个名称为 usertable 的数据表,具体表结构如表 1-1 所示。

表 1-1 usertable 数据表

字 段 名 称	类　　型	中 文 含 义
username	Character varying(10)	登录用户名
password	Character varying(10)	登录密码
type	int	用户类型:1 表示管理员,2 表示普通用户

3. 在表 usertable 添加 3 条记录,如表 1-2 所示。

表 1-2 usertable 中的记录

登录用户名	密　　码	用 户 类 型
zjut	zjut	1
admin	admin	1
temp	temp	2

4. 修改工程 javaweb-prj1。添加 PostgreSQL 驱动程序库文件 postgresql-8.4-702.jdbc4.jar 到工程中。

5. 修改工程 javaweb-prj1。创建数据库操作类 UserDAO.java,实现按用户名和密码校验用户是否合法的功能。具体代码如下所示。

```
package cn.edu.zjut.dao;
import java.sql.*;import javax.sql.*;import javax.naming.*;
import cn.edu.zjut.model.UserBean;
public class UserDAO{
    private static final String GET_ONE_SQL =
        "SELECT * FROM usertable WHERE username = ? and password = ?";
    public UserDAO(){}
    public Connection getConnection(){
        Connection conn = null;
        String driver = "org.postgresql.Driver";
        String dburl = "jdbc:postgresql://127.0.0.1:5432/myDB";
        String username = "dbuser";          //数据库登录用户名
        String password = "dbpassword";      //数据库登录密码
        try{
            Class.forName(driver);           //加载数据库驱动程序
```

```
                conn = DriverManager.getConnection(dburl,username,password);
            }catch( Exception e ){ e.printStackTrace(); }
            return conn;
    }
    public boolean searchUser(UserBean user){
        //按用户名和密码校验用户是否合法
        Connection conn = null;
        PreparedStatement pstmt = null;
        ResultSet rst = null;
        try{
            conn = getConnection();
            pstmt = conn.prepareStatement(GET_ONE_SQL);
            pstmt.setString(1, user.getUsername());
            pstmt.setString(2, user.getPassword());
            rst = pstmt.executeQuery();
            if(rst.next()){
               return true;
            }
        }catch(SQLException se){
            se.printStackTrace();
            return false;
        }finally{
            try{
                 pstmt.close();
                 conn.close();
            }catch(SQLException se){ se.printStackTrace(); }
        }
        return false;
    }
}
```

6. 修改工程 javaweb-prj1 中的控制器 LoginController.java。

（1）引入 UserDAO 类，具体代码如下所示。

⋮
```
import cn.edu.zjut.dao.UserDAO;
```
⋮

（2）重写 checkUser(UserBean user)方法。实现通过数据库表 usertable 的记录来校验用户名和密码合法性的功能，具体修改代码如下所示。

⋮
```
boolean checkUser(UserBean user){
        UserDAO ud = new UserDAO();
            if( ud.searchUser(user) ) {
              return true;
            }
            return false;
}
```
⋮

7. 将 javaweb-prj1 工程重新部署在 Tomcat 服务器上。

8. 运行 login.jsp 页面，输入用户名和密码并运行。观察运行结果。如果出现运行错误,请尝试修正。

9. 对 javaweb-prj1 工程进行如下修改。

(1) 将登录成功的条件修改为:用户名、密码和用户类型三者匹配。

(2) 新增一个用户注册视图 register.jsp,用于普通用户的注册。

(3) 修改 UserDAO 类。新增一个 insert(UserBean user)方法,用于向 usertable 表中插入一条记录。

(4) 运行并观察结果。

四、实验要求

1. 填写并上交实验报告,报告中应包括如下内容。

(1) 运行结果截图。

(2) 查找相关资料,总结 DAO 设计模型的优点。

(3) 实验收获及总结。

2. 上交程序源代码,代码中应有相关注释。

第二部分　轻量级Java EE编程技术

- 实验二　Struts2基础应用
  ——基于Struts2框架的用户登录模块
- 实验三　Struts2的控制器组件Action
  ——登录用户的功能扩展
- 实验四　Struts2的工作流程
  ——登录用户的高级功能
- 实验五　Hibernate基础应用
  ——基于Hibernate框架的用户登录模块
- 实验六　Hibernate的体系结构
  ——登录用户信息的增、删、改、查
- 实验七　Hibernate关联关系映射
  ——登录用户的地址管理
- 实验八　SSH整合(Spring4+Struts2+Hibernate4)
  ——基于SSH的用户注册模块
- 实验九　Spring的核心机制：控制反转(IoC)
  ——登录用户的购物车
- 实验十　Spring的面向切面编程(AOP)
  ——用户登录模块的增强处理

实验二

Struts2 基础应用——基于 Struts2 框架的用户登录模块

基础实验——Struts2 框架搭建

一、实验目的

1. 掌握 Struts2 应用的基本开发步骤和常规配置。
2. 观察表单参数与 Action 属性的赋值关系,观察 Action 的 execute()方法及其返回值,并能够正确应用。
3. 观察配置文件 struts.xml 中的主要元素及属性,并能够正确应用。
4. 理解 Struts2 框架中 MVC 设计模式的体现,理解 Action、FilterDispatcher 和 struts.xml 的主要作用,并能够正确使用。

二、基本知识与原理

1. Struts2 是从 WebWork 框架上发展起来的 MVC 框架。
2. FilterDispatcher 是 Struts2 中的核心控制器。客户端对服务器端的请求将被 FilterDispatcher 过滤。若请求需要调用某个 Action,则框架将根据配置文件 struts.xml,找到需要调用的 Action 类。
3. Action 类是一个符合一定命名规范的 Java SE 类,作为业务控制器使用。Action 中的 execute()方法用于调用 Model 层的业务逻辑类,并根据返回结果确定页面导航。
4. 若 Action 类中需要使用表单提交的请求参数,那么必须在 Action 类中声明与表单域的名字对应的变量,并为变量提供 getters/setters 方法。
5. Action 类需要在 struts.xml 中进行配置才能使用。
6. 编译运行基于 Struts2 框架的 Web 工程。需要导入 Struts2 的 8 个核心 JAR 包,如表 2-1 所示。

表 2-1 Struts2 的 8 个核心 JAR 包

文件名	说明
struts2-core-2.3.15.1.jar	Struts2 框架的核心类库
xwork-core-2.3.15.1.jar	XWork 类库,Struts2 的构建基础
Ognl-3.0.6.jar	Struts2 使用的一种表达式语言类库

续表

文 件 名	说 明
freemarker-2.3.19.jar	Struts2 的标签模板使用类库
javassist-3.11.0.GA.jar	代码生成工具包
commons-lang3-3.1.jar	Apache 语言包，是 java.lang 包的扩展
commons-io-2.0.1.jar	Apache IO 包
commons-fileupload-1.3.jar	Struts2 文件上传依赖包

三、实验内容及步骤

1. 登录 http://struts.apache.org/download.cgi 站点，下载最新版的 Struts2(Full Distribution)。

2. 在 Eclipse 中新建 Web 工程 struts-prj1。

3. 将 Struts2 中的 8 个核心包增加到 Web 应用中，即复制到%workspace%struts-prj1\WebContent\WEB-INF\lib 路径下，如图 2-1 所示。

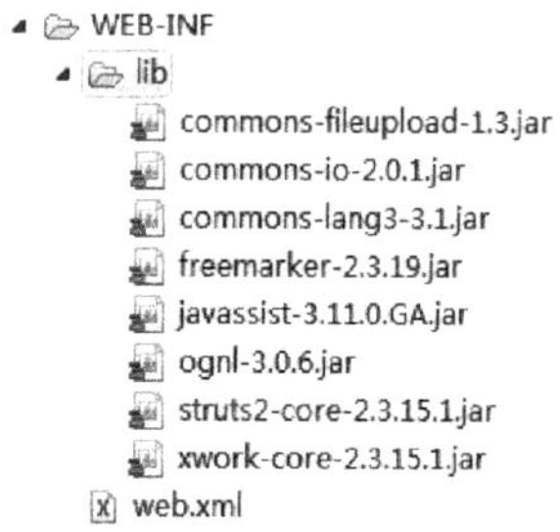

图 2-1 Struts2 的 8 个核心包

4. 在 struts-prj1 中新建 login.jsp 页面，作为用户登录的视图代码片段如下所示。注意表单 form 中 action 属性的值，以及用户名与密码输入框中 name 属性的值。

```
<form action = "login" method = "post">
        请输入用户名: < input name = "loginUser.account"
type = "text">< BR >
        请输入密码: < input name = "loginUser.password" type = "password">
        < input type = "submit" value = "登录">
</form >
```

5. 在 struts-prj1 中新建 loginSuccess.jsp 和 loginFail.jsp 页面，分别作为登录成功和登录失败的视图，并在页面中显示“登录成功”和“登录失败”的提示信息。

6. 在 struts-prj1 中新建 cn.edu.zjut.bean 包，并在其中创建 UserBean.java，用于记录用户信息，代码片断如下所示。注意该 JavaBean 中属性名的写法。

```
package cn.edu.zjut.bean;
public class UserBean {
    private String account = "";
    private String password = "";

    public String getAccount() {
        return account;
    }
    public void setAccount(String account) {
        this.account = account;
    }
    public String getPassword() {
        return password;
    }
```

```
        public void setPassword(String password) {
            this.password = password;
        }
    }
```

7. 在 struts-prj1 中新建 cn.edu.zjut.service 包,并在其中创建 UserService.java,用于实现登录逻辑。为简化登录逻辑,将登录成功的条件设置为:用户名和密码相同。代码片断如下所示。

```
package cn.edu.zjut.service;
import cn.edu.zjut.bean.UserBean;
public class UserService {
    public boolean login(UserBean loginUser) {
        if (loginUser.getAccount().equals(loginUser.getPassword())) {
            return true;
        }
        return false;
    }
}
```

8. 在 struts-prj1 中新建 cn.edu.zjut.action 包,并在其中创建 UserAction.java。调用登录逻辑,并根据登录结果的不同而返回不同的内容,代码片断如下所示。注意该 Action 中的属性名及相应的 getters 和 setters 方法、execute()方法及返回值。

```
package cn.edu.zjut.action;
import cn.edu.zjut.bean.UserBean;
import cn.edu.zjut.service.UserService;
public class UserAction {
    private UserBean loginUser;
    public UserBean getLoginUser() {
        return loginUser;
    }
    public void setLoginUser(UserBean loginUser) {
        this.loginUser = loginUser;
    }
    public String execute() {
        UserService userServ = new UserService();
        if (userServ.login(loginUser)) {
            return "success";
        }
        return "fail";
    }
}
```

9. 在工程 struts-prj1 的 src 目录中创建 struts.xml 文件,用于配置 Action 并设置页面导航,代码片段如下所示。注意 action 标签中 name 属性和 class 属性的值,以及 result 子标签的属性的值。

```
<?xml version = "1.0" encoding = "UTF-8" ?>
```

```xml
<!DOCTYPE struts PUBLIC
    "-//Apache Software Foundation//DTD Struts Configuration 2.3//EN"
    "http://struts.apache.org/dtds/struts-2.3.dtd">
<struts>
  <package name="strutsBean" extends="struts-default" namespace="/">
        <action name="login" class="cn.edu.zjut.action.UserAction">
            <result name="success">/loginSuccess.jsp</result>
            <result name="fail">/loginFail.jsp</result>
        </action>
  </package>
</struts>
```

10. 编辑 Web 应用的 web.xml 文件,增加对 Struts2 核心 Filter 的配置,代码片段如下所示。

```xml
<web-app>
⋮
<!-- 定义 Struts2 的核心 Filter -->
<filter>
    <filter-name>struts2</filter-name>
    <filter-class>
      org.apache.struts2.dispatcher.ng.filter
                    .StrutsPrepareAndExecuteFilter
    </filter-class>
</filter>
<!-- 让 Struts2 的核心 Filter 拦截所有请求 -->
<filter-mapping>
    <filter-name>struts2</filter-name>
    <url-pattern>/*</url-pattern>
</filter-mapping>
⋮
</web-app>
```

11. 将 struts-prj1 部署在 Tomcat 服务器上。

12. 通过浏览器访问 login.jsp 页面,并记录运行结果。

13. 尝试对步骤 4、6、8、9 中的关键代码进行修改。观察修改后的运行结果。

四、实验要求

1. 填写并上交实验报告,报告中应包括如下内容。

(1) 运行结果截图、修改后的关键代码及相应的运行结果或报错信息。

(2) 根据实验过程,总结 jsp 页面、Action 类、Service 类、JavaBean、Filter 和 struts.xml 文件的作用,整理 Struts2 应用中从请求到响应的完整流程,思考并总结 Struts2 框架中 MVC 的体现。

(3) 根据实验过程,总结表单参数与 Action 属性的赋值关系,并记录下来。

(4) 根据实验过程,总结 Action 的 execute() 方法的作用和特点,并记录下来。

(5) 根据实验过程,查找相关资料,写出本实验中配置文件 struts.xml 里各元素及其属性的作用。

(6) 碰到的问题及解决方案或对问题的思考。
(7) 实验收获及总结。
2. 上交程序源代码,代码中应有相关注释。

提高实验——Struts2 标签

一、实验目的

1. 进一步熟悉 Struts2 应用的基本开发步骤和常规配置。
2. 进一步熟悉 Action 及配置文件 struts.xml 的应用方法。
3. 掌握 Struts2 标签的基本使用方法。
4. 能熟练使用 Struts2 的常用标签,能参考 Struts2 标签的使用说明文档,对各类标签进行灵活应用。

二、基本知识与原理

1. 使用 Struts2 标签的形式来表达页面逻辑。可以尽量避免在视图中使用 Java 代码,让逻辑与显示分离,提高视图的可维护性。

2. Struts2 标签库的主要 tld 文件为 struts-tags.tld,位于 struts2-core-2.3.15.1.jar 包中。另一个与 Ajax 相关的标签库 tld 文件为 struts-dojo-tags.tld,位于 struts2-dojo-plugin-2.3.15.1.jar 包中。

3. Struts2 标签的使用步骤和使用 JSTL 相同,只需在 JSP 页面中使用 taglib 指令引入标签库中 tld 文件的 uri,并指定前缀即可。例如：<%@ taglib prefix="s" uri="/struts-tags"%>。

4. 根据 Struts2 标签的主要作用,可以将其分为:用于生成页面元素的 UI 标签、用于实现流程控制的控制类标签、用于控制数据的数据标签和用于支持 Ajax 的标签。

三、实验内容及步骤

1. 为了能够使用与 Ajax 相关的标签,要将 Struts2 中的 struts2-dojo-plugin-2.3.15.1.jar 包增加到工程 struts-prj1 中,即复制到%workspace%struts-prj1\WebContent\WEB-INF\lib 路径下,然后刷新工程。

2. 在工程 struts-prj1 中增加用户注册功能。新建 register.jsp 页面作为用户注册的视图。页面使用 Struts2 的 UI 标签来生成表单元素,包括用户名、密码、确认密码、真实姓名、性别、生日、联系地址、联系电话和电子邮箱,代码片段如下所示。

```
<%@ taglib prefix="s" uri="/struts-tags"%>
<%@ taglib prefix="sx" uri="/struts-dojo-tags"%>
<html>
<head><sx:head/></head>
<body>
<s:form action="register" method="post">
    <s:textfield name="loginUser.account" label="请输入用户名"/>
    <s:password name="loginUser.password" label="请输入密码"/>
```

```
        ⋮
    <s:radio name = "loginUser.sex" list = "#{1 : '男', 0 : '女'}" label
        = "请输入性别"/>
    <sx:datetimepicker name = "loginUser.birthday" displayFormat
        = "yyyy-mm-dd" label = "请输入生日"/>
        ⋮
    <s:submit value = "注册"/>
    <s:reset value = "重置"/>
</s:form>
</body>
</html>
```

3. 在 struts-prj1 中新建 regFail.jsp 页面,作为注册失败的视图,并在页面中显示"注册失败"。

4. 在 struts-prj1 中新建 regSuccess.jsp 页面,作为注册成功的视图。使用 Struts2 的数据标签和控制标签来生成注册成功的信息,并将登录用户信息保存在会话范围内。代码片段如下所示。

```
<!-- 数据标签 property -->
<s:property value = "loginUser.name"/>
<!-- 控制标签 if/else -->
<s:if test = "loginUser.sex">
    <s:text name = "先生,"/>
</s:if>
<s:else>
    <s:text name = "女士,"/>
</s:else>
您注册成功了!
<!-- 数据标签 set -->
<s:set name = "user" value = "loginUser" scope = "session"/>
```

5. 修改 UserBean.java。增加属性,用于记录注册的用户信息。代码片段如下所示。

```
public class UserBean {
    private String account = "";
    private String password = "";
    private String repassword = "";
    private String name = "";
    private String sex = "";
    private String birthday = "";
    private String address = "";
    private String phone = "";
    private String email = "";
    //省略 getters/setters 方法
}
```

6. 修改 UserService.java。增加用户注册逻辑。为简化注册逻辑,将注册成功的条件设置为:用户名、密码和确认密码相同,而且不为空字符串。

7. 修改 UserAction.java 中的 execute()方法,参照基础实验部分写入代码,用于调用注册逻辑,并根据注册状态返回相应的内容。

8. 修改 struts.xml 文件,对用户注册信息进行配置并设置页面导航。

9. 将 struts-prj1 重新部署在 Tomcat 服务器上。通过浏览器访问 register.jsp 页面,并记录运行结果。

10. 查找 Struts2 标签的相关资料,并尝试将其他的 Struts2 标签应用在实验步骤 2、3、4 中,观察应用后的运行结果。

四、实验要求

1. 填写并上交实验报告,报告中应包括如下内容。
（1）运行结果截图。
（2）应用各种 Struts2 标签的关键代码及相应的运行结果或报错信息。
（3）分析实验过程,查找相关资料,总结 Struts2 中标签及其属性的作用和用法,并记录下来。
（4）碰到的问题及解决方案或对问题的思考。
（5）实验收获及总结。
2. 上交程序源代码,代码中应有相关注释。

扩展实验——Struts2 的国际化

一、实验目的

1. 进一步熟悉 Struts2 标签的基本使用方法。
2. 能使用 Struts2 标签实现国际化。
3. 了解配置文件 struts.properties 以及国际化资源文件的作用和基本使用方法。

二、基本知识与原理

1. 通过将不同语言版本的字符保存在属性文件中,Struts2 的国际化机制在不修改程序主体的前提下,就能实现其不同语言版本的应用。

2. 在 Web 应用中选择需要进行国际化的内容。不是在页面中直接输出该信息,而是通过 Struts2 标签输出一个键值。该键值在不同语言环境下对应不同的字符串。例如"＜s:textfield name="loginUser.account" lable="请输入用户名"/＞"代码中的"请输入用户名"是需要国际化的内容,将以其键值代替为"＜s:textfield name="loginUser.account" key="login.account.lable"/＞"。

3. 需要进行国际化的内容将以键值对(key＝value)的形式写入 Struts2 的国际化资源文件中,如"login.account.lable＝请输入用户名"。该资源文件名可以自定义,但是后缀必须是 properties。资源文件应放在 Web 应用的类加载路径下。每一个语言版本都需要创建一个资源文件。

4. 通过 Struts2 的配置文件 struts.properties 来配置资源文件的基础名。若资源文件的基础名为 message,则 message_zh_CN.properties 是对应的中文资源文件,message_en_US.properties 是对应的美国英语资源文件。

5. 浏览器将根据其默认的语言版本，自动调用相应语言的资源文件，从而在页面中展示不同的语言效果。

三、实验内容及步骤

1. 在工程 struts-prj1 的 src 目录下新建一个 cn.edu.zjut.local 包，把所有的资源文件放置其中，如 message_zh_CN.properties、message_en_US.properties 等。

2. 将工程 struts-prj1 中用户登录模块的 3 个 JSP 页面进行国际化处理。选择需要进行国际化的内容，以键值对的形式写入资源文件中。代码片段如下所示。

```
# message_en_US.properties
login.account.lable = Please input your account
login.password.lable = Please input your password
login.submit.button = submit

# message_zh_CN.properties
login.account.lable = 请输入用户名
login.password.lable = 请输入密码
login.submit.button = 登录
```

3. 使用 JDK 中的 native2ASCII 工具，将 message_zh_CN.properties 重新编码，将中文字符都转化为 unicode 码，从而避免乱码问题，如图 2-2 所示。

```
F:\eclipse-workspace\struts-prj1\src\local>native2ASCII message_zh_CN_temp.properties message_zh_CN.properties
F:\eclipse-workspace\struts-prj1\src\local>
```

图 2-2　使用 native2ASCII 工具

具体代码如下所示。

```
# message_zh_CN.properties
login.account.lable = \u8bf7\u8f93\u5165\u7528\u6237\u540d
login.password.lable = \u8bf7\u8f93\u5165\u5bc6\u7801
login.submit.button = \u767b\u5f55
```

4. 在工程 struts-prj1 的 src 目录中创建 struts.properties 文件，通过它加载资源文件。具体代码如下所示。

```
struts.custom.i18n.resources = cn.edu.zjut.local.message
struts.i18n.encoding = GBK
```

或者在 JSP 页面中临时加载资源文件，代码片段如下所示。

```
<s:i18n name = "cn.edu.zjut.local.message">
<s:form action = "login" method = "post"> … </s:form>
</s:i18n>
```

5. 修改 login.jsp、loginSuccess.jsp 和 loginFail.jsp 页面，通过 Struts2 标签实现国际化。代码片段如下所示。

```
<!-- login.jsp -->
<s:textfield name = "loginUser.account" key = "login.account.lable"/>
<s:password name = "loginUser.password"
key = "login.password.lable"/>
<s:submit name = "submit" key = "login.submit.button"/>
```

6. 将 struts-prj1 重新部署在 Tomcat 服务器上。

7. 设置浏览器的语言首选项,如图 2-3 所示。通过浏览器访问 login.jsp 页面,观察并记录运行结果。

图 2-3 浏览器的语言首选项

8. 尝试修改用户注册模块,实现其国际化,并记录运行结果。

四、实验要求

1. 填写并上交实验报告,报告中应包括如下内容。
(1) 运行结果截图。
(2) 应用国际化的关键代码及相应的运行结果或报错信息。
(3) 查找相关资料,总结配置文件 struts.properties 的作用并记录下来。
(4) 碰到的问题及解决方案或对问题的思考。
(5) 实验收获及总结。
2. 上交程序源代码,代码中应有相关注释。

实验三

基础实验——Action 的自定义方法

一、实验目的

1. 掌握 Struts2 的 Action 类中自定义方法的使用。
2. 掌握 Struts2 中 Action 类的不同调用方式和相应的配置方法。
3. 掌握 Action 的实例化情况,理解 Action 与 Servlet 在实例化上的区别。
4. 理解 JSP 文件中获取 Action 属性的过程。
5. 了解 Struts2 支持的 Action 处理结束后的结果类型。

二、基本知识与原理

1. Action 类中的默认方法名为 execute(),它可以被自动调用。

2. 在 Action 中也允许定义其他方法名。可以同时定义多个方法,来分别处理不同的逻辑。

3. 如果 Action 中使用了自定义方法,则该 Action 就需要特定的配置。一般有以下 4 种调用方式。

(1) 在 struts.xml 文件中通过 method 属性指定方法名。
(2) 使用动态方法调用方式(DMI)。
(3) 使用提交按钮的 method 属性。
(4) 使用通配符配置 Action。

4. Action 类是多实例的,Action 类的属性是线程安全的。

5. 在 JSP 页面中,可以通过 Struts2 标签调用 Action 中对应的 getter 方法,从而输出 Action 的属性值。

6. 一个 Action 处理完用户请求后,将返回一个字符串作为逻辑视图名,再通过 struts.xml 文件中的配置将逻辑视图名与物理视图资源关联起来。Struts2 默认提供了一系列的结果类型(struts-default.xml 配置文件的 result-types 标签里列出了所支持的结果类型)。结果类型决定了 Action 处理结束后,将调用哪种视图资源来呈现处理结果。

三、实验内容及步骤

1. 在 Eclipse 中新建 Web 工程 struts-prj2，并将 Struts2 中的 8 个核心包添加到工程中。

2. 在 struts-prj2 中新建 login.jsp 页面，作为用户登录的视图；新建 loginFail.jsp 页面，作为登录失败的视图；新建 loginSuccess.jsp 页面，作为登录成功的视图（可重用实验二中基础实验里的页面代码）。

3. 在 struts-prj2 中新建 register.jsp 页面，作为用户注册的视图；新建 regFail.jsp 页面，作为注册失败的视图；新建 regSuccess.jsp 页面，作为注册成功的视图（可重用实验二中提高实验里的页面代码）。

4. 在 struts-prj2 中新建 cn.edu.zjut.bean 包，并在其中创建 UserBean.java，用于记录用户信息（可重用实验二中提高实验里的 UserBean.java 代码）。

5. 在 struts-prj2 中新建 cn.edu.zjut.service 包，并在其中创建 UserService.java，用于实现登录逻辑和注册逻辑（可重用实验二中提高实验里的 UserService.java 代码）。

6. 在 struts-prj2 中新建 cn.edu.zjut.action 包，并在其中创建 UserAction.java，定义 login()方法和 register()方法，参照实验二写入代码，分别用于调用登录逻辑和注册逻辑。代码片段如下所示。

```java
package cn.edu.zjut.action;
    ⋮
public class UserAction {
    ⋮
    public String login() {
        UserService userServ = new UserService();
        if (userServ.login(loginUser)) {
            return "success"; }
        return "fail";
    }
    public String register() {
        UserService userServ = new UserService();
        if (userServ.register(loginUser)) {
            return "registersuccess"; }
        return "registerfail";
    }
}
```

7. 在工程 struts-prj2 的 src 目录中创建 struts.xml 文件，用于配置 Action 并设置页面导航。通过 action 标签中 method 属性指定方法名，代码片段如下所示。

```xml
<struts>
  <package name = "strutsBean" extends = "struts-default" namespace = "/">
    <action name = "login" class = "cn.edu.zjut.action.UserAction"
            method = "login">
      <result name = "success">/loginSuccess.jsp</result>
      <result name = "fail">/loginFail.jsp</result>
    </action>
```

```xml
<action name="register" class="cn.edu.zjut.action.UserAction"
        method="register">
    <result name="registersuccess">/regSuccess.jsp</result>
    <result name="registerfail">/regFail.jsp</result>
</action>
    </package>
</struts>
```

8. 编辑 Web 应用的 web.xml 文件,增加 Struts2 核心 Filter 的配置。

9. 将 struts-prj2 部署在 Tomcat 服务器上。通过浏览器访问 login.jsp 与 register.jsp 页面,并记录运行结果。

10. 查找相关资料,尝试使用 Action 自定义方法的其他 3 种调用和配置方式:动态方法调用方式(DMI)、提交按钮的 method 属性、通配符配置 Action。记录关键配置和运行结果。

11. 修改 UserAction.java。增加 UserAction 类的构造方法 UserAction();增加 count 属性,用于测试 Action 的实例化情况,代码片段如下所示。

```java
public class UserAction {
    private Integer count = 0;
    public UserAction(){
        System.out.println("创建了一个 UserAction 类对象.");
    }
    public Integer getCount() {
        return count;
    }
    public String login() {
        count++;                                  //Action 实例化情况测试
        UserService userServ = new UserService();
        if (userServ.login(loginUser)) {
            return "success";
        }
        return "fail";
    }
    ...
}
```

12. 修改 loginSuccess.jsp,在页面中使用<s:property>标签输出 Action 中的 count 值。

13. 重新将 struts-prj2 部署在 Tomcat 服务器上。通过浏览器访问 login.jsp 页面,并进行多次刷新,记录运行结果。

14. 修改 struts.xml 文件,将 UserAction 的页面导航设置为 redirect 结果类型。代码片段如下所示。

```xml
<struts>
    <package name="strutsBean" extends="struts-default" namespace="/">
        <action name="login" class="cn.edu.zjut.action.UserAction"
                method="login">
            <result name="success" type="redirect">
```

```
            /loginSuccess.jsp</result>
      <result name="fail">/loginFail.jsp</result>
    </action>
  </package>
</struts>
```

15. 重新将 struts-prj2 部署在 Tomcat 服务器上。通过浏览器访问 login.jsp 页面，观察登录成功后 loginSuccess.jsp 页面的输出，并记录下来。

四、实验要求

1. 填写并上交实验报告，报告中应包括如下内容。
（1）运行结果截图。
（2）查找相关资料，根据实验过程，总结 Action 自定义方法的 4 种调用和配置方式。
（3）根据实验过程，记录实验步骤 11、12 中修改后的关键代码以及相应的运行结果或报错信息。分析 Action 的实例化情况，将 Action 与 Servlet 在实例化情况上进行对比，并记录下来。
（4）根据实验步骤 11、12，查找相关资料，分析 JSP 文件中获取 Action 属性的主要过程，并记录下来。
（5）根据实验步骤 11~15，观察两次 loginSuccess.jsp 页面输出上的区别，分析原因并记录下来。
（6）解压缩 Struts2 的核心包 struts2-core-2.3.15.1.jar，找到 struts-default.xml 配置文件，在其中的 result-types 标签里列出了 Struts2 所支持的结果类型。查找相关资料，总结这些结果类型的作用和特点，并记录下来。
（7）碰到的问题及解决方案或对问题的思考。
（8）实验收获及总结。

2. 上交程序源代码，代码中应有相关注释。

提高实验——ActionSupport 与输入校验

一、实验目的

1. 了解 Action 接口的作用，理解 ActionSupport 类的作用。
2. 掌握在 Struts2 中使用校验器或手工编码的方式以及对请求参数进行数据校验的方法，掌握在 JSP 页面中显示错误信息和提示信息的方法。
3. 掌握在 Action 中使用国际化资源文件的方法。
4. 掌握 Struts2 内置类型转换器的作用和使用方法。

二、基本知识与原理

1. 为了让用户开发的 Action 类更规范，Struts2 提供了一个 Action 接口。该接口定义了 Struts2 的 Action 处理类应该实现的规范。
2. Struts2 还为 Action 接口提供了一个实现类——ActionSupport。该类提供了若干默认方法,包括:默认的处理用户请求的方法(execute()方法)、数据校验的方法、添加校验错误信息的方法、获取国际化信息的方法等。表3-1中列出了部分重要方法。

表 3-1　ActionSupport 类的部分重要方法

方　法　名	说　　明
public String execute()	默认的处理用户请求的方法,直接返回 SUCCESS 字符串
public void validate()	空的输入校验方法,常被 Action 类覆盖,用于实现对输入参数的校验
public void addActionError(String anErrorMessage)	将 Action 级别的错误信息添加到 Action 中
public void addActionMessage(String aMessage)	将 Action 级别的消息添加到 Action 中
public void addFieldError(String fieldName, String errorMessage)	将域级错误信息添加到特定的域中
public String getText(String aTextName)	从国际化资源文件中获取属性值,其中的参数是属性文件的 key 值

3. Struts2 框架提供了校验器和手工编码两种方式来对请求参数进行数据校验。当 Action 类继承了 ActionSupport 类时,就可以通过定义名为 <ActionClassName>-<ActionAliasName>-validation.xml 的校验规则文件的方法进行校验器校验,也可以通过重写 ActionSupport 类的 validate()方法或 validateXxx()方法进行手动校验。

4. 在 JSP 页面中,使用 Struts2 标签生成的表单,能将域级别的错误信息自动显示到表单元素处。

5. 在 JSP 页面中使用 fielderror 标签,可以集中显示所有的域级错误信息;使用 actionerror 标签,可以显示所有的 Action 级别错误信息;使用 actionmessage 标签,可以显示 Action 消息。

6. Struts2 框架中提供了部分内置的类型转换器。可以将请求参数的 String 类型转换成基本数据类型及对应的包装器类型、日期类型、数组类型、集合类型等。当 Action 类继承了 ActionSupport 类时,内置的类型转换器将默认生效,可以直接被使用。

7. 如需修改默认的类型转换校验信息,则要在 Action 类的包中声明名为"Action 类名.properties"的局部属性文件。

8. Struts2 框架同时还支持自定义类型转换器,可将请求参数转换成任意一种类型。

三、实验内容及步骤

1. 在 struts-prj2 中修改 UserAction 类,使其继承 ActionSupport 类,并在 UserAction 类中覆盖 ActionSupport 类的 validate()方法,用于对用户登录的请求参数 account 和 password 的校验。若用户名或密码为空,则使用 addFieldError(域级)添加错误信息。代码片段如下所示。

```
import com.opensymphony.xwork2.ActionSupport; …
public class UserAction extends ActionSupport {
    ⋮
    public void validate() {
```

```
        String account = loginUser.getAccount();
        String pwd = loginUser.getPassword();
        if (account == null || account.equals("")) {
            this.addFieldError("loginUser.account", "请输入您的用户名!");
        }
        if (pwd == null || pwd.equals("")) {
            this.addFieldError("loginUser.password", "请输入您的密码!");
        }
    }
}
```

2. 修改 struts.xml 文件,在 Action 的配置中增加 validate()方法校验出错时的页面导航(<result name="input">),代码片段如下所示。

```
<struts>
    <package name="strutsBean" extends="struts-default" namespace="/">
        <action name="login" class="cn.edu.zjut.action.UserAction"
                method="login">
            <result name="success">/loginSuccess.jsp</result>
            <result name="fail">/loginFail.jsp</result>
            <result name="input">/login.jsp</result>
        </action>
        ⋮
    </package>
</struts>
```

3. 重新将 struts-prj2 部署在 Tomcat 服务器上。通过浏览器访问 login.jsp 页面,观察并记录运行结果。

4. 修改 login.jsp 页面。在表单前增加 fielderror 标签<s:fielderror/>,再通过浏览器访问 login.jsp 页面。观察并记录运行结果。

5. 修改 UserAction.java。在调用登录逻辑的 login()方法中,对登录情况进行校验。若登录成功,使用 addActionMessage()方法添加"登录成功!"的 Action 提示消息;若登录失败,使用 addActionError()方法添加 Action 级别的错误信息。代码片段如下所示。

```
public class UserAction extends ActionSupport {
    ⋮
    public String login() {
        UserService userServ = new UserService();
        if (userServ.login(loginUser)) {
            this.addActionMessage("登录成功!");
            return "success";
        } else {
            this.addActionError("用户名或密码错误,请重新输入!");
            return "fail";
        }
    }
}
```

6. 修改 login.jsp 页面,增加 actionerror 标签(<s:actionerror/>)显示 Action 级别的错误信息。修改 loginSuccess.jsp 页面,使用 actionmessage 标签(<s:actionmessage/>)

显示 Action 提示消息。

7. 修改 struts.xml 文件中用户登录的页面导航设置,将登录失败时转向的页面从 loginFail.jsp 修改为 login.jsp。

8. 重新将 struts-prj2 部署在 Tomcat 服务器上。通过浏览器访问 login.jsp 页面,观察并记录运行结果。

9. 在工程 struts-prj2 中创建 UserAction-login-validation.xml 校验规则文件,使其与 UserAction 类位于同一目录下。配置校验信息,使用校验器对请求参数进行校验(代码如下)。

```xml
<?xml version="1.0" encoding="UTF-8"?>
<!DOCTYPE validators PUBLIC
    "-//Apache Struts//XWork Validator 1.0.2//EN"
    "http://struts.apache.org/dtds/xwork-validator-1.0.2.dtd">

<validators>
    <field name="loginUser.account">
        <field-validator type="requiredstring">
            <param name="trim">true</param>
            <message>用户名不能为空</message>
        </field-validator>
    </field>
    <field name="loginUser.password">
        <field-validator type="requiredstring">
            <param name="trim">true</param>
            <message>密码不能为空</message>
        </field-validator>
    </field>
</validators>
```

10. 重新将 struts-prj2 部署在 Tomcat 服务器上。通过浏览器访问 login.jsp 页面,观察并记录运行结果。

11. 参考实验二中的扩展实验,将 login.jsp、loginSuccess.jsp、loginFail.jsp 3 个页面进行国际化处理,把需要进行国际化的内容以键值对的形式写入资源文件 message_zh_CN.properties 和 message_en_US.properties 中。

12. 在资源文件中添加校验信息的键值对,并使用 native2ASCII 工具,将 message_zh_CN.properties 重新编码,将中文字符都转化为 unicode 码,代码片段如下所示。

```
#message_en_US.properties
#field error message
login.account.null = Please input your account!
login.password.null = Please input your password!
#action error message
login.error = Account or password error, please input again!
#action message
login.success = Login successfully!
```

13. 在工程 struts-prj2 的 src 目录中创建 struts.properties 文件,通过它加载资源文

件。具体代码如下所示。

```
struts.custom.i18n.resources = cn.edu.zjut.local.message
struts.i18n.encoding = GBK
```

14. 修改 UserAction.java，使用 ActionSupport 类的 getText() 方法，获取国际化资源文件中的信息。代码片段如下所示。

```java
public class UserAction extends ActionSupport {
    ⋮
    public String login() {
        UserService userServ = new UserService();
        if (userServ.login(loginUser)) {
            this.addActionMessage(this.getText("login.success"));
            return "success";
        } else {
            this.addActionError(this.getText("login.error"));
            return "fail";
        }
    }
    public void validate() {
        String account = loginUser.getAccount();
        String pwd = loginUser.getPassword();
        if (account == null || account.equals("")) {
            this.addFieldError("loginUser.account",
                    this.getText("login.account.null"));
        }
        if (pwd == null || pwd.equals("")) {
            this.addFieldError("loginUser.password",
                    this.getText("login.password.null"));
        }
    }
}
```

15. 修改 UserAction-login-validation.xml，获取国际化资源文件中的信息。代码片段如下所示。

```xml
<validators>
    <field name = "loginUser.account">
        <field-validator type = "requiredstring">
            <param name = "trim">true</param>
            <message key = "login.account.null"/>
        </field-validator>
    </field>
    ⋮
</validators>
```

16. 重新将 struts-prj2 部署在 Tomcat 服务器上。通过浏览器访问 login.jsp 页面，观察并记录运行结果。

17. 修改 UserBean.java。将用于保存注册用户生日的变量类型改为 Date 类型，使用

Struts2 内置的类型转换器对请求参数进行校验。

18. 重新将 struts-prj2 部署在 Tomcat 服务器上。通过浏览器访问 register.jsp 页面。当用户输入的生日不合法时，观察并记录运行结果。

19. 在工程 struts-prj2 的 cn.edu.zjut.action 包中创建局部属性文件 UserAction.properties，修改类型转换的校验信息，并使用 native2ASCII 工具将 UserAction.properties 重新编码。代码片段如下所示。

```
#其中,invalid.fieldvalue不能随意改,loginUser.birthday是请求参数域名,
#应根据实际需要进行修改
invalid.fieldvalue.loginUser.birthday = 生日必须是日期,并符合"yyyy-mm-dd"格式
```

20. 重新将 struts-prj2 部署在 Tomcat 服务器上。通过浏览器访问 register.jsp 页面。当用户输入的生日不合法时，观察并记录运行结果。

21. 参考实验步骤 9，在工程 struts-prj2 的 cn.edu.zjut.action 包中创建 UserAction-register-validation.xml 文件。增加校验信息的配置，使用校验器对用户注册的请求参数进行校验，要求注册时两次密码输入相同、email 地址格式符合要求等。

22. 重新将 struts-prj2 部署在 Tomcat 服务器上。通过浏览器访问 register.jsp 页面，观察并记录运行结果。

23. 修改 UserAction 类，将 validate() 的方法名称改为 validateLogin()，并增加 validateRegister() 方法。参考实验步骤 1，使用手工编码方式对请求参数进行数据校验。

24. 重新将 struts-prj2 部署在 Tomcat 服务器上。通过浏览器访问 register.jsp 页面，观察并记录运行结果。

四、实验要求

1. 填写并上交实验报告，报告中应包括如下内容。

(1) 运行结果截图。

(2) 分析实验步骤 1~7 和步骤 23~24，查找相关资料，总结 Action 类中 validate() 方法和 validateXxx() 方法的作用、使用时的要点或注意事项，总结在 JSP 页面中显示错误信息和提示信息的方法，并记录实验中相应的关键代码。

(3) 分析实验步骤 8~9、21~22，总结使用校验器进行校验的方法。在 Struts2 的核心包 xwork-core-2.3.15.1.jar\com\opensymphony\xwork2\validator\validators 路径下找到 default.xml 文件。查找相关资料，总结校验规则文件中主要元素的作用和配置方法，并结合相应案例将其记录下来。

(4) 根据实验步骤 10~16，总结在 Action 中使用国际化资源文件的步骤及方法，并结合实验中相应的关键代码将其记录下来。

(5) 根据实验步骤 17~20，总结 Struts2 中常用的内置类型转换器及其使用方法。

(6) 碰到的问题及解决方案或对问题的思考。

(7) 实验收获及总结。

2. 上交程序源代码，代码中应有相关注释。

扩展实验——Action 类与 Servlet API

一、实验目的

1. 掌握在 Action 中访问 Servlet API 的 4 种方法,理解这 4 种方法的区别。
2. 进一步理解 Action 与 Servlet 的区别。
3. 进一步熟悉 Struts2 标签的使用方法。

二、基本知识与原理

1. Struts2 框架中的 Action 类没有与任何 Servlet API 耦合,因此 Action 类可以脱离 Servlet 容器环境进行单元测试。

2. ActionContext 是 com.opensymphony.xwork2 包中的一个类。该类表示一个 Action 运行时的上下文。

3. 当 Action 类需要通过请求、会话或上下文存取属性时,可以通过 ActionContext 类完成,也可以通过实现 Struts2 提供的接口 RequestAware、SessionAware 和 ApplicationAware 来完成,而不必调用 Servlet API 中的 HttpServletRequest、HttpSession 和 ServletContext 对象,从而保持 Action 与 Servlet API 的解耦状态。

4. 要在 Action 类中直接访问 Servlet API,可以通过实现 Struts2 提供的接口 ServletContextAware、ServletRequestAware、ServletResponseAware 来完成,也可以通过 ServletActionContext 工具类实现,但 Action 将与 Servlet API 直接耦合。

三、实验内容及步骤

1. 在 struts-prj2 中修改 UserAction 类,可通过 ActionContext 获取请求、会话和上下文对象相关联的 Map 对象来实现存取属性的功能。代码片段如下所示。

```
public class UserAction extends ActionSupport {
    ⋮
  private Map request,session,application;
  public String login() {
        //获取 ActionContext 对象
        ActionContext ctx = ActionContext.getContext();
        //通过 ActionContext 对象获取请求、会话和上下文对象相关联的 Map 对象
        request = (Map) ctx.get("request");
        session = (Map) ctx.getSession();
        application = (Map) ctx.getApplication();
        //访问 application 范围的属性值
        Integer counter = (Integer)application.get("counter");
        if(counter == null)
            counter = 1;
        else
            counter = counter + 1;
        //设置 application 范围的属性
        application.put("counter", counter);
```

```java
        UserService userServ = new UserService();
        if (userServ.login(loginUser)) {
            //设置session范围的属性
            session.put("user", loginUser.getAccount());
            //设置request范围的属性
            request.put("tip", "您已登录成功");
            return "success";
        } else {
            return "fail";
        }
    }
}
```

2. 修改loginSuccess.jsp页面，从请求、会话和上下文对象中获取属性值并显示在页面中。代码片段如下所示。

```
<body>
本站访问次数为:<s:property value="#application.counter"/><br>
<s:property value="#session.user"/>,
<s:property value="#request.tip"/>
</body>
```

3. 重新将struts-prj2部署在Tomcat服务器上。通过浏览器访问login.jsp页面，观察并记录运行结果。

4. 修改UserAction类，通过实现Struts2提供的接口RequestAware、SessionAware和ApplicationAware，来获取请求、会话和上下文对象相关联的Map对象，从而实现存取属性的功能。代码片段如下所示。

```java
public class UserAction extends ActionSupport
        implements RequestAware,SessionAware,ApplicationAware{
    ⋮
    private Map request,session,application;
    public void setRequest(Map<String, Object> request) {
        this.request = request;
    }
    public void setSession(Map<String, Object> session) {
        this.session = session;
    }
    public void setApplication(Map<String, Object> application) {
        this.application = application;
    }
    public String login() {
        Integer counter = (Integer)application.get("counter");
        if(counter == null)
            counter = 1;
        else
            counter = counter + 1;
        application.put("counter", counter);
        UserService userServ = new UserService();
        if (userServ.login(loginUser)) {
```

```
                session.put("user", loginUser.getAccount());
                request.put("tip", "您已登录成功");
                return "success";
            } else {
                return "fail";
            }
        }
    }
```

5. 重新将 struts-prj2 部署在 Tomcat 服务器上。通过浏览器访问 login.jsp 页面,观察并记录运行结果。

6. 修改 UserAction 类,查找相关资料,尝试通过 ServletActionContext 工具类直接访问 Servlet API,实现与上述实验步骤 1～3 相同的功能,重新运行并记录结果。

7. 修改 UserAction 类,查找相关资料,尝试通过接口 ServletContextAware、ServletRequestAware、ServletResponseAware 直接访问 Servlet API,实现与上述实验步骤 4～5 相同的功能,重新运行并记录结果。

8. 尝试利用 Servlet API 添加购物车功能,在工程 struts-prj2 的 cn.edu.zjut.bean 包中创建 Item.java,用于记录商品信息。代码片段如下所示。

```
public class Item {
        private String itemID;
        private String name;
        private String description;
        private double cost;

    public Item(String itemID, String name,
            String description, double cost) {
            setItemID(itemID);
            setTitle(name);
            setDescription(description);
            setCost(cost);
        }
        //省略 getters/setters 方法
}
```

9. 在工程 struts-prj2 的 cn.edu.zjut.bean 包中创建 ItemOrder.java,用于记录购物车中的一条商品信息和购买数量。代码片段如下所示。

```
public class ItemOrder {
    private Item item;
    private int numItems;

    public ItemOrder(Item item) {
        setItem(item);
        setNumItems(1);
    }
    //省略 getters/setters 方法
}
```

10. 在工程 struts-prj2 的 cn.edu.zjut.bean 包中创建 ShopppingCart.java，用于记录用户的购物车信息。为简化操作，在购物车的构造函数中加入商品信息。代码片段如下所示。

```java
public class ShoppingCart {
    private List itemsOrdered;
    public ShoppingCart() {
        itemsOrdered = new ArrayList();
        Item item = new Item("book001", "Java EE 技术实验指导教程",
                "Web 程序设计知识回顾、" + "轻量级 Java EE 应用框架、"
                + "企业级 EJB 组件编程技术、" + "Java EE 综合应用开发.", 19.95);
        ItemOrder itemorder = new ItemOrder(item);
        itemorder.setNumItems(2);
        itemsOrdered.add(itemorder);
    }
    public List getItemsOrdered() {
        return (itemsOrdered);
    }
    public synchronized void addItem(String itemID) {}
    public synchronized void setNumOrdered(String itemID,
        int numOrdered) {}
}
```

11. 修改 UserAction 类的 login 方法。当登录成功时，创建 ShoppingCart 对象并存入会话中。

12. 修改 loginSuccess.jsp 页面，从会话中获取购物车信息并显示在该页面中。代码片段如下所示。

```
<body>
<table border=1>
<s:iterator value="#session.shoppingcart.itemsOrdered">
<tr>
<th>编号</th><th>名称</th><th>说明</th><th>单价</th><th>数量</th>
</tr>
<tr>
    <td><s:property value="item.itemID"/></td>
    <td><s:property value="item.name"/></td>
    <td><s:property value="item.description"/></td>
    <td><s:property value="item.cost"/></td>
    <td><s:property value="numItems"/></td>
</tr>
</s:iterator>
</table>
</body>
```

13. 重新将 struts-prj2 部署在 Tomcat 服务器上。通过浏览器访问 login.jsp 页面，观察并记录运行结果。

四、实验要求

1. 填写并上交实验报告，报告中应包括如下内容。

（1）运行结果截图。

（2）根据实验过程，观察 Action 中访问 Servlet API 的 4 种方法，总结这 4 种方法的区别，并记录下来。

（3）整理购物车功能相关的关键代码，记录相应的运行结果或报错信息。

（4）碰到的问题及解决方案或对问题的思考。

（5）实验收获及总结。

2．上交程序源代码，代码中应有相关注释。

实验四

Struts2 的工作流程
——登录用户的高级功能

基础实验——拦截器与过滤器

一、实验目的

1. 掌握 Struts2 自定义拦截器的基本开发步骤和配置方法。
2. 掌握 Struts2 自定义过滤器的基本开发步骤和配置方法。
3. 理解拦截器和过滤器的特点和区别。
4. 了解 Struts2 默认拦截器栈中包含的主要拦截功能。
5. 深入理解 Struts2 的工作原理和基本工作过程。

二、基本知识与原理

1. Struts2 的控制器主要由 3 个层次组成,分别是过滤器、拦截器和业务控制器 Action。

2. 过滤器是 Struts2 控制器的最前端控制器。要使用过滤器就需要在 web.xml 中进行配置。FilterDispatcher 是 Struts2 应用中必须进行配置和使用的过滤器。该过滤器的主要功能包括执行 Action、清空 ActionContext 对象等。

3. 拦截器是 Struts2 中第二个层次的控制器,它能够在 Action 执行前后完成一些通用功能。

4. Struts2 内建了大量拦截器。这些拦截器以 name-class 对的形式配置在 struts-default.xml 文件中。如果 struts.xml 中定义的 package 继承了 Struts2 默认的 struts-default 包,就可以直接使用默认拦截器栈 defaultStack。

5. Struts2 也允许自定义拦截器。自定义拦截器类须实现 Interceptor 接口,并覆盖接口中的 intercept 方法,用于实现拦截器的主要功能。自定义拦截器须在 struts.xml 中进行配置才能使用。

6. 若在 struts.xml 中为 Action 指定了一个拦截器,则默认拦截器栈 defaultStack 将会失效。为了继续使用默认拦截器,必须将其进行显式的配置。

三、实验内容及步骤

1. 在 Eclipse 中新建 Web 工程 struts-prj3,并将 Struts2 中的 8 个核心包添加到工

实验四 Struts2 的工作流程——登录用户的高级功能

程中。

2. 在 struts-prj3 中新建 login.jsp 页面，作为用户登录的视图；新建 loginSuccess.jsp 页面，作为登录成功的视图（可重用实验三中基础实验里的页面代码）。

3. 在 struts-prj3 中新建 cn.edu.zjut.bean 包，并在其中创建 UserBean.java，用于记录用户信息（可重用实验三中基础实验里的 UserBean.java 代码）。

4. 在 struts-prj3 中新建 cn.edu.zjut.service 包，并在其中创建 UserService.java，用于实现登录逻辑和注册逻辑（可重用实验三中基础实验里的 UserService.java 代码）。

5. 在 struts-prj3 中新建 cn.edu.zjut.action 包，并在其中创建 UserAction.java。定义 login()方法，用于调用登录逻辑（可重用实验三中基础实验里的 UserAction.java 代码）。

6. 在 struts-prj3 的 cn.edu.zjut.bean 包中创建 Item.java，用于记录商品信息（可重用实验三中扩展实验里的 Item.java 代码）。

7. 在 struts-prj3 的 cn.edu.zjut.service 包中创建 ItemService.java，用于获取所有商品信息。为简化操作，将商品信息直接写入代码中。代码片段如下所示。

```java
package cn.edu.zjut.service;
...
public class ItemService {
    public List getAllItems(){
        List items = new ArrayList();
        items.add(new Item("book001", "Java EE 技术实验指导教程",
            "Web 程序设计知识回顾、" + "轻量级 Java EE 应用框架、"
            + "企业级 EJB 组件编程技术、" + "Java EE 综合应用开发.", 19.95));
        items.add(new Item("book002", "Java EE 技术",
            "Struts 框架、Hibernate 框架、Spring 框架、"
            + "会话 Bean、实体 Bean、消息驱动 Bean", 29.95));
        return items;
    }
}
```

8. 在 struts-prj3 的 cn.edu.zjut.action 包中创建 ItemAction.java。定义 execute()方法，用于调用"获取所有商品信息"逻辑。代码片段如下所示。

```java
package cn.edu.zjut.service;
...
public class ItemAction extends ActionSupport {
    private List items;
    //省略 getters/setters 方法

    public String getAllItems() {
        ItemService itemServ = new ItemService();
        items = itemServ.getAllItems();
        System.out.println("ItemAction excuted!");
        return "success";
    }
}
```

9. 在 struts-prj3 中创建 itemList.jsp 页面，作为显示所有商品信息的视图，代码片段如下所示。

```
<body>
<center>商品列表</center>
<table border=1>
<tr>
<th>编号</th><th>书名</th><th>说明</th><th>单价</th>
</tr>
<s:iterator value="items">
<tr>
    <td><s:property value="itemID"/></td>
    <td><s:property value="name"/></td>
    <td><s:property value="description"/></td>
    <td><s:property value="cost"/></td>
</tr>
</s:iterator>
</table>
</body>
```

10. 修改 loginSuccess.jsp 页面,作为登录成功的视图,并在视图中增加超链接,用于查看所有商品信息,代码片段如下所示。

```
<a href="./allItems">查看所有商品信息</a>
```

11. 在工程 struts-prj2 的 src 目录中创建 struts.xml 文件,用于配置 Action 并设置页面导航,代码片段如下所示。

```
<struts>
  <package name="strutsBean" extends="struts-default" namespace="/">
    <action name="login" class="cn.edu.zjut.action.UserAction"
                         method="login">
        <result name="success">/loginSuccess.jsp</result>
        <result name="fail">/login.jsp</result>
    </action>
    <action name="allItems" class="cn.edu.zjut.action.ItemAction"
            method="getAllItems">
        <result name="success">/itemList.jsp</result>
    </action>
  </package>
</struts>
```

12. 编辑 Web 应用的 web.xml 文件,增加 Struts2 核心 Filter 的配置。

13. 将 struts-prj3 部署在 Tomcat 服务器上。通过浏览器访问 login.jsp。登录成功后单击超链接,查看所有商品信息。观察并记录运行结果。

14. 在 struts-prj3 中新建 cn.edu.zjut.interceptors 包,并在其中创建拦截器 AuthorityInterceptor.java,用于实现用户权限控制功能,使得只有登录用户才有查看所有商品信息的权限。代码片段如下所示。

```
package cn.edu.zjut.interceptors;
import java.util.Map;
```

```java
import com.opensymphony.xwork2.Action;
import com.opensymphony.xwork2.ActionContext;
import com.opensymphony.xwork2.ActionInvocation;
import com.opensymphony.xwork2.interceptor.AbstractInterceptor;
public class AuthorityInterceptor extends AbstractInterceptor{
    public String intercept(ActionInvocation invocation)
                throws Exception {
        System.out.println("Authority Interceptor executed!");
        ActionContext ctx = invocation.getInvocationContext();
        Map session = ctx.getSession();
        String user = (String)session.get("user");
        if(user!= null){
            return invocation.invoke();
        }
        else{
            ctx.put("tip", "您还没有登录,请输入用户名和密码登录系统");
            return Action.LOGIN;
        }
    }
}
```

15. 修改 UserAction.java,通过 ActionContext 获取与 session 对象相关联的 Map 对象。当用户登录成功时,将用户名作为属性放入 session 范围内。代码片段如下所示。

```java
public class UserAction extends ActionSupport {
    ⋮
  private Map session;
  public String login() {
        ActionContext ctx = ActionContext.getContext();
        session = (Map) ctx.getSession();
        UserService userServ = new UserService();
        if (userServ.login(loginUser)) {
            session.put("user", loginUser.getAccount());
            return "success";
        } else {
            return "fail";
        }
    }
}
```

16. 修改 struts.xml 文件,增加拦截器的配置。代码片段如下所示。

```xml
<package name = "strutsBean" extends = "struts-default" namespace = "/">
    <!-- 定义一个名为 authority 的拦截器 -->
    <interceptors>
        <interceptor name = "authority"
            class = "cn.edu.zjut.interceptors.AuthorityInterceptor"/>
    </interceptors>
    <action name = "allItems" class = "cn.edu.zjut.action.ItemAction"
            method = "getAllItems">
        <result name = "login">/login.jsp</result>
```

```
        <result name = "success">/itemList.jsp</result>
        <!-- 配置系统默认拦截器 -->
        <interceptor - ref name = "defaultStack"/>
        <!-- 配置 authority 拦截器 -->
        <interceptor - ref name = "authority"/>
    </action>
      ⋮
</package>
```

17. 重新将 struts-prj3 部署在 Tomcat 服务器上。首先,不经用户登录,直接通过浏览器访问 loginSuccess.jsp 页面。单击超链接,查看所有商品信息,观察并记录运行结果。然后,访问 login.jsp 页面,经用户登录后进入 loginSuccess.jsp 页面。单击超链接,查看所有商品信息,观察并记录运行结果。

18. 在 struts-prj3 中新建 cn.edu.zjut.filters 包,并在其中创建过滤器 AccessFilter.java,用于实现 JSP 页面的过滤功能,使得只有登录用户才能查看除 login.jsp 和 register.jsp 之外的其他 JSP 页面。代码片段如下所示。

```java
package cn.edu.zjut.filters;
public class AccessFilter implements Filter {
    ⋮
    public void doFilter(ServletRequest arg0, ServletResponse arg1,
        FilterChain filterChain) throws IOException, ServletException
    {
            System.out.println("Access Filter executed!");
        HttpServletRequest request = (HttpServletRequest)arg0;
        HttpServletResponse response = (HttpServletResponse)arg1;
        HttpSession session = request.getSession();
        if(session.getAttribute("user") == null &&
            request.getRequestURI().indexOf("login.jsp") == -1 &&request.getRequestURI()
.indexOf("register.jsp") == -1){
            response.sendRedirect("login.jsp");
            return ;
         }
          filterChain.doFilter(arg0, arg1);
     }
}
```

19. 修改 web.xml 文件,增加过滤器的配置。代码片段如下所示。

```xml
<web - app>
    ⋮
    <filter>
         <filter - name>accessFilter</filter - name>
<filter - class>cn.edu.zjut.filters.AccessFilter</filter - class>
    </filter>
    <filter - mapping>
         <filter - name>accessFilter</filter - name>
             <url - pattern>*.jsp</url - pattern>
    </filter - mapping>
</web - app>
```

20. 重新将 struts-prj3 部署在 Tomcat 服务器上。首先，不经用户登录直接，通过浏览器访问 loginSuccess.jsp 和 itemList.jsp 页面，观察并记录运行结果。然后，访问 login.jsp 页面，经用户登录后进入 loginSuccess.jsp 页面。单击超链接，查看所有商品信息，观察并记录运行结果。

四、实验要求

1. 填写并上交实验报告，报告中应包括如下内容。

（1）运行结果截图。

（2）根据实验过程，查找相关资料，整理自定义拦截器类的作用和实现方法，整理 Interceptor 接口中 intercept(ActionInvocation inv)、init() 和 destroy() 方法的作用，并记录下来。记录实验步骤 14 中 intercept(ActionInvocation inv) 方法返回值的含义。

（3）根据实验过程，查找相关资料，整理自定义拦截器的配置步骤、注意事项，并记录配置文件中相关标签的作用。

（4）在 Struts2 核心包 struts2-core-2.3.15.1.jar 的 struts-default.xml 文件中找到 struts-default 包默认的拦截器栈 defaultStack 的定义，查找相关资料，整理该拦截器栈中包含的主要拦截功能。

（5）根据实验过程，查找相关资料，整理自定义过滤器的实现方法和配置步骤。将拦截器与过滤器进行比较，并将两者的特点及区别记录下来。

（6）根据实验过程，查找相关资料，总结 Struts2 的工作原理和基本工作过程，并记录下来。

（7）碰到的问题及解决方案或对问题的思考。

（8）实验收获及总结。

2. 上交程序源代码，代码中应有相关注释。

提高实验——值栈与 OGNL

一、实验目的

1. 理解值栈的概念，了解值栈接口的主要方法和使用步骤。
2. 掌握使用 OGNL 获取值栈内容的方法。
3. 掌握使用 OGNL 获取 session、application 等其他对象的方法。

二、基本知识与原理

1. Struts API 中的 com.opensymphony.xwork2.util.ValueStack 被称为值栈。值栈是一个数据区域，该区域中保存了应用范围内的所有数据和 Action 处理的用户请求数据。

2. 值栈被存储在 ActionContext 对象中，因此可以在任何节点访问其中的内容。

3. ValueStack 接口中的主要方法有：Object findValue(String expr)，可以通过表达式查找值栈中对应的值；void setValue(String expr, Object value)，用于将对象及其表达式存储到值栈中。

4. OGNL(Object Graphic Navigation Language),即对象图导航语言。它是 Struts 默认的表达式语言。

5. OGNL 基础单位称为导航链。一个基本的链由属性名、方法调用、数组或集合元素组成。

6. 在 Struts2 中,值栈是 OGNL 上下文的根对象,可以直接对其进行访问,而 application、session 等其他对象不是根对象,需要使用#进行访问。

三、实验内容及步骤

1. 修改 itemList.jsp 页面,通过值栈对象获得属性。

(1) 修改 itemList.jsp 页面的 page 指令,导入相关的 java 包。代码片段如下所示。

```
<%@ page language = "java" contentType = "text/html; charset = GB18030"
    pageEncoding = "GB18030"
    import = "com.opensymphony.xwork2.util.ValueStack,
             java.util.List,java.util.Iterator,
             cn.edu.zjut.bean.Item" %>
```

(2) 修改<body></body>标签中的代码,获得值栈对象,通过值栈接口的 findValue 方法获得值栈中对象的值并输出。代码片段如下所示。

```
<body>
<%
    ValueStack vs = (ValueStack)request.
                        getAttribute("struts.valueStack");
    List itemList = (List)vs.findValue("items");
%>
<center>商品列表</center>
<table border = 1>
<tr>
<th>编号</th>
<th>书名</th>
<th>说明</th>
<th>单价</th>
</tr>
<% Iterator it = itemList.iterator();
   while(it.hasNext()){
       Item item = (Item)it.next(); %>
<tr>
    <td><% = item.getItemID() %></td>
    <td><% = item.getTitle() %></td>
    <td><% = item.getDescription() %></td>
    <td><% = item.getCost() %></td>
</tr>
<% } %>
</table>
</body>
```

2. 重新访问 login.jsp 页面。登录成功后,单击超链接,查看所有商品信息,观察并记

实验四 Struts2 的工作流程——登录用户的高级功能

录运行结果。

3. 修改 itemList.jsp 页面。通过 OGNL 获得值栈内容。由于值栈是 OGNL 上下文的根对象，所以可以直接访问。代码片段如下所示。

```
<body>
<center>商品列表</center>
<table border = 1>
<tr>
<th>编号</th><th>书名</th><th>说明</th><th>单价</th>
</tr>
<s:iterator value = "items">
<tr>
    <td><s:property value = "itemID"/></td>
    <td><s:property value = "name"/></td>
    <td><s:property value = "description"/></td>
    <td><s:property value = "cost"/></td>
</tr>
</s:iterator>
</table>
</body>
```

4. 修改 itemList.jsp 页面。通过 OGNL 访问 session 对象。由于 session、application 等对象不是根对象，所以需要使用 ♯ 进行访问。代码片段如下所示。

```
<body>
<s:property value = "♯session.user"/>,您好!
<center>商品列表</center>
  ⋮
</body>
```

5. 重新访问 login.jsp 页面。登录成功后，单击超链接，查看所有商品信息，观察并记录运行结果。

6. 修改 itemList.jsp 页面。使用符号 ♯ 过滤集合，取出价格小于 20 元的商品和名称为"Java EE 技术实验指导教程"的商品。代码片段如下所示。

```
价格低于 20 元的商品有:<br>
<s:iterator value = "items.{?♯this.cost<20}">
    <li><s:property value = "title"/>:
        <s:property value = "cost" />元</li>
</s:iterator>
<p>
名称为《Java EE 技术实验指导教程》的商品的价格为:
<s:property value = "items.
       {?♯this.title == 'Java EE 技术实验指导教程'}.{cost}[0]"/>元
```

7. 修改 itemList.jsp 页面。使用符号 % 计算 OGNL 表达式的值，比较使用 % 和不使用 % 的输出情况。代码片段如下所示。

```
<s:url value = "items.{title}[0]"/><br>
<s:url value = "%{items.{title}[0]}"/>
```

8. 重新访问 login.jsp 页面。登录成功后，单击超链接，查看所有商品信息，观察并记录运行结果。

四、实验要求

1. 填写并上交实验报告，报告中应包括如下内容。
（1）运行结果截图。
（2）根据实验步骤 1~2，查找相关资料，整理 ValueStack 接口及其主要方法的作用和开发步骤，并记录下来。
（3）根据实验步骤 3~5，查找相关资料，整理 OGNL 可访问的对象和基本语法，并记录下来。
（4）根据实验步骤 6~8，查找相关资料，整理 OGNL 的 3 种常用符号#、%和$的作用及使用方法，并记录下来。
（5）碰到的问题及解决方案或对问题的思考。
（6）实验收获及总结。
2. 上交程序源代码，代码中应有相关注释。

扩展实验——Struts2 的异常处理

一、实验目的

1. 掌握 Struts2 应用中处理异常的方式。
2. 掌握在 struts.xml 中对 Action 类配置异常映射的方法。

二、基本知识与原理

1. 在 Struts2 应用中要使用 Action 来调用 Model，因此 Struts2 应用中的异常在 Model 层抛出后，通常在 Action 类中进行处理。
2. Action 可以直接使用 try/catch 捕获异常，然后返回结果视图，并跳转到相关页面处理异常。
3. 抛出异常后，也可以不在 Action 类中捕获，而使用 throws 声明异常，交给 Struts2 框架处理。
4. Struts2 允许通过 struts.xml 文件来配置异常的处理，使用<exception-mapping>标签声明异常映射，指定发生该类型异常时跳转的结果视图。

三、实验内容及步骤

1. 在 struts-prj3 中新建 cn.edu.zjut.exception 包，并在其中创建自定义异常类 UserException.java，代码片断如下所示。

```
package cn.edu.zjut.exception;
public class UserException extends Exception{
    public UserException() { super(); }
    public UserException(String msg) { super(msg); }
```

实验四　Struts2的工作流程——登录用户的高级功能

```java
    public UserException(String msg, Throwable cause) {
        super(msg, cause);
    }
    public UserException(Throwable cause) { super(cause); }
}
```

2. 在 struts-prj3 中新建 loginException.jsp 页面，作为用户登录异常的视图，代码片段如下所示。

`<body>登录异常!</body>`

3. 修改 UserService.java。在 login 方法中，当用户名为 admin 时将抛出自定义异常，当密码包含 and 或 or 时将抛出 SQLException 异常。代码片段如下所示。

```java
public class UserService {
    public boolean login(UserBean loginUser) throws Exception{
        if (loginUser.getAccount().equalsIgnoreCase("admin")){
            throw new UserException("用户名不能为admin");
        }
        if (loginUser.getPassword().toUpperCase().contains(" AND ")
            || loginUser.getPassword().toUpperCase().contains(" OR ")){
            throw new java.sql.SQLException("密码不能包括'and '或' or '");
        }
        if (loginUser.getAccount().equals(loginUser.getPassword())) {
            return true;
        }
        else
            return false;
    }
    ⋮
}
```

4. 修改 UserAction.java。在 login 方法中使用 try/catch 捕获异常，并在捕获异常后返回结果视图，跳转到相关页面。代码片段如下所示。

```java
public class UserAction extends ActionSupport {
    ⋮
    public String login(){
        ⋮
        UserService userServ = new UserService();
        try {
            if (userServ.login(loginUser)) {
                ⋮
                return "success";
            } else {
                ⋮
                return "fail";
            }
        } catch (Exception e) {
            e.printStackTrace();
```

```
            return "exception";
        }
    }
}
```

5. 修改 struts.xml 文件,设置异常页面导航,代码片段如下所示。

```xml
<struts>
  <package name = "strutsBean" extends = "struts-default" namespace = "/">
    <interceptors>
        <interceptor name = "authority"
            class = "interceptors.AuthorityInterceptor"/>
    </interceptors>
    <action name = "login" class = "cn.edu.zjut.action.UserAction"
                        method = "login">
        <result name = "success">/loginSuccess.jsp</result>
        <result name = "fail">/login.jsp</result>
        <result name = "exception">/loginException.jsp</result>
    </action>
    ⋮
  </package>
</struts>
```

6. 将 struts-prj3 重新部署在 Tomcat 服务器上。通过浏览器访问 login.jsp,尝试错误登录,观察并记录运行结果。

7. 修改 UserAction.java。在 login 方法中抛出异常而不捕获,将异常交给 Struts2 框架处理,代码片段如下所示。

```java
public class UserAction extends ActionSupport {
    ⋮
    public String login() throws Exception {
        ⋮
        UserService userServ = new UserService();
        try {
            if (userServ.login(loginUser)) {
                ⋮
                return "success";
            } else {
                ⋮
                return "fail";
            }
        } catch (Exception e) {
            throw e;
        }
    }
}
```

8. 修改 struts.xml 文件,使用<exception-mapping>标签完成异常配置,并通过全局和局部两种方式进行异常映射,代码片段如下所示。

```xml
<struts>
  <package name="strutsBean" extends="struts-default" namespace="/">
    <interceptors>
        <interceptor name="authority"
            class="interceptors.AuthorityInterceptor"/>
    </interceptors>
    <global-results>
        <result name="sqlExcp">/loginException.jsp</result>
    </global-results>
    <global-exception-mappings>
        <exception-mapping exception="java.sql.SQLException"
                                result="sqlExcp"/>
    </global-exception-mappings>
    <action name="login" class="cn.edu.zjut.action.UserAction"
                        method="login">
        <exception-mapping result="userExcp"
                exception="cn.edu.zjut.exception.UserException"/>
        <result name="userExcp">/loginException.jsp</result>
        <result name="success">/loginSuccess.jsp</result>
        <result name="fail">/login.jsp</result>
    </action>
    ⋮
  </package>
</struts>
```

9. 修改 loginException.jsp 页面,使用 Struts2 标签输出异常信息,代码片段如下所示。

```
<body>
异常信息:<s:property value="exception.message"/>
</body>
```

10. 将 struts-prj3 重新部署在 Tomcat 服务器上。通过浏览器访问 login.jsp 页面,尝试错误登录,观察并记录运行结果。

四、实验要求

1. 填写并上交实验报告,报告中应包括如下内容。
(1) 运行结果截图。
(2) 根据实验步骤1~6,查找相关资料,整理自定义异常类的方法和步骤,并记录下来。
(3) 根据实验步骤1~6,将 Action 使用 try/catch 捕获异常并返回结果视图的关键代码和相关配置记录下来。
(4) 根据实验步骤 7~10,查找相关资料,整理 Struts2 框架处理异常的机制,整理 struts.xml 文件配置异常映射的方法以及相关标签的作用,并记录下来。
(5) 碰到的问题及解决方案或对问题的思考。
(6) 实验收获及总结。
2. 上交程序源代码,代码中应有相关注释。

实验五

Hibernate 基础应用——基于 Hibernate 框架的用户登录模块

基础实验——Hibernate 框架搭建

一、实验目的

1. 掌握 Hibernate 开发环境搭建的基本步骤。

2. 观察持久化类与数据库表的映射关系,观察相应的 Hibernate 映射文件(.hbm.xml 文件)配置,并能够做简单应用。

3. 观察 Hibernate 配置文件(hibernate.cfg.xml)中的主要元素及属性配置,并能够做简单应用。

二、基本知识与原理

1. Hibernate 是一个 ORM(Object-Relational Mapping)框架,用于把对象模型表示的对象映射到基于 SQL 的关系模型数据结构中去,采用完全面向对象的方式来操作数据库。

2. Hibernate 的主要作用是简化应用的数据持久层编程。它不仅能管理 Java 类到数据库表的映射,还能提供数据查询和获取数据的方法,从而大幅减少了开发人员编写 SQL 和 JDBC 代码的时间。

3. Hibernate 框架主要包括持久化对象(Persistent Objects)、Hibernate 配置文件(一般被命名为 *.cfg.xml)、Hibernate 映射文件(一般被命名为 *.hbm.xml)三部分。

4. 编译、运行基于 Hibernate 框架的工程,需要导入相应的 Hibernate 类库。

5. 由于 Hibernate 底层是基于 JDBC 的,因此在应用程序中使用 Hibernate 执行持久化操作时,也需要导入相关的 JDBC 驱动(例如 MySQL 数据库驱动)。

三、实验内容及步骤

1. 登录 http://dev.mysql.com/downloads/mysql/ 站点,下载并安装 MySQL 数据库。

2. 在 MySQL 中创建一个名称为 hibernatedb 的数据库,并在该数据库中创建一个名称为 customer 的数据表,表结构如表 5-1 所示。

3. 在表 customer 中添加 3 条记录,具体内容如表 5-2 所示。

4. 登录 http://downloads.mysql.com/archives/c-j/ 站点,下载 MySQL JDBC 驱动。

实验五　Hibernate 基础应用——基于 Hibernate 框架的用户登录模块

表 5-1　customer 数据表

字 段 名 称	类　　型	中 文 含 义
customerID	INTEGER(11)，Primary key，Not Null	用户编号
account	VARCHAR(20)	登录用户名
password	VARCHAR(20)	登录密码
name	VARCHAR(20)	真实姓名
sex	BOOLEAN(1)	性别
birthday	DATE	出生日期
phone	VARCHAR(20)	联系电话
email	VARCHAR(100)	电子邮箱
address	VARCHAR(200)	联系地址
zipcode	VARCHAR(10)	邮政编码
fax	VARCHAR(20)	传真号码

表 5-2　customer 中的记录

用 户 编 号	登录用户名	登录密码
1	zjut	zjut
2	admin	admin
3	temp	temp

5. 在 Eclipse 中新建 Web 工程 hibernate-prj1，并添加 MySQL 驱动程序库文件和 Struts2 核心包到工程中。

6. 登录 http://www.hibernate.org/downloads 站点，下载 Hibernate 发布版（如：hibernate-release-4.3.5.Final），并将其解压缩。将 Hibernate 发布版中 lib\required 里的 jar 包添加到工程 hibernate-prj1 中。

7. 在 http://commons.apache.org/proper/commons-logging/download_logging.cgi 站点，下载 commons-logging-1.2-bin.zip，并将其解压缩。将 commons-logging-1.2.jar 添加到工程 hibernate-prj1 中。

8. 如图 5-1 所示，在工程 hibernate-prj1 中新建 Hibernate 相关文件，用于将对象模型表示的对象映射到关系模型数据结构中，从而采用完全面向对象的方式来操作数据库。

图 5-1　Hibernate 相关文件

（1）其中，cn.edu.zjut.po 包中的 Customer.java 是与 customer 数据库表相对应的持久化类。持久化类通常用 POJO 编程模式实现，代码片段如下所示。

```
package cn.edu.zjut.po;

import java.util.Date;
public class Customer implements java.io.Serializable {

    private int customerId;
    private String account; private String password;
```

```java
        private String name; private Boolean sex;
        private Date birthday; private String phone;
        private String email; private String address;
        private String zipcode; private String fax;

        public Customer() {
        }

        public Customer(int customerId) {
            this.customerId = customerId;
        }

        public Customer(int customerId, String account, String password,
            String name, Boolean sex, Date birthday, String phone,
            String email, String address, String zipcode, String fax) {
            this.customerId = customerId;
            this.account = account; this.password = password;
            this.name = name; this.sex = sex;
            this.birthday = birthday; this.phone = phone;
            this.email = email; this.address = address;
            this.zipcode = zipcode; this.fax = fax;
        }
        //省略 getters/setters 方法
}
```

(2) Customer.hbm.xml 是 Hibernate 映射文件。其中名为＜class＞的元素表示持久化类 Customer.java 与数据库表 customer 的映射关系，其子元素＜id＞表示持久化类中的主键，子元素＜property＞表示持久化类中的其他属性与数据库表中某列的映射关系。具体代码如下所示。

```xml
<?xml version = "1.0"?>
<!DOCTYPE hibernate-mapping PUBLIC "-//Hibernate/Hibernate Mapping DTD 3.0//EN"
"http://hibernate.sourceforge.net/hibernate-mapping-3.0.dtd">
<hibernate-mapping>
    <class name = "cn.edu.zjut.po.Customer" table = "customer" catalog = "hibernatedb">
        <id name = "customerId" type = "int">
            <column name = "customerID" />
            <generator class = "assigned" />
        </id>
        <property name = "account" type = "string">
            <column name = "account" length = "20" unique = "true" />
        </property>
        <property name = "password" type = "string">
            <column name = "password" length = "20" />
        </property>
        <property name = "name" type = "string">
            <column name = "name" length = "20" />
        </property>
        <property name = "sex" type = "java.lang.Boolean">
            <column name = "sex" />
```

```xml
        </property>
        <property name = "birthday" type = "date">
            <column name = "birthday" length = "10" />
        </property>
        <property name = "phone" type = "string">
            <column name = "phone" length = "20" />
        </property>
        <property name = "email" type = "string">
            <column name = "email" length = "100" />
        </property>
        <property name = "address" type = "string">
            <column name = "address" length = "200" />
        </property>
        <property name = "zipcode" type = "string">
            <column name = "zipcode" length = "10" />
        </property>
        <property name = "fax" type = "string">
            <column name = "fax" length = "20" />
        </property>
    </class>
</hibernate-mapping>
```

（3）hibernate.cfg.xml 是 Hibernate 配置文件，用于设置 JDBC 连接相关的属性，如连接数据库的地址、用户名、密码等，并在其中增加 Customer.hbm.xml 映射文件声明。具体代码如下所示。

```xml
<?xml version = "1.0" encoding = "utf-8"?>
<!DOCTYPE hibernate-configuration PUBLIC
"-//Hibernate/Hibernate Configuration DTD 3.0//EN"
"http://hibernate.sourceforge.net/hibernate-configuration-3.0.dtd">
<hibernate-configuration>
    <session-factory name = "HibernateSessionFactory">
        <property name = "hibernate.connection.driver_class">
                com.mysql.jdbc.Driver</property>
        <property name = "hibernate.connection.url">
jdbc:mysql://localhost:3306/hibernatedb</property>
        <property name = "hibernate.connection.username">
                root</property>
        <property name = "hibernate.connection.password"/>
        <property name = "hibernate.dialect">
                org.hibernate.dialect.MySQLDialect</property>
        <mapping resource = "cn/edu/zjut/po/Customer.hbm.xml" />
    </session-factory>
</hibernate-configuration>
```

9. 在 hibernate-prj1 中新建 login.jsp 页面，作为用户登录的视图；新建 loginSuccess.jsp 页面，作为登录成功的视图（可重用实验二中基础实验里的页面代码）。

10. 在 hibernate-prj1 中新建 cn.edu.zjut.dao 包，并在其中创建数据库操作类 CustomerDAO.java。具体代码如下所示。

```java
package cn.edu.zjut.dao;
import java.util.List;
import org.hibernate.Query;
import org.hibernate.SessionFactory;
import org.hibernate.Session;
import org.hibernate.cfg.Configuration;
import org.apache.commons.logging.Log;
import org.apache.commons.logging.LogFactory;

public class CustomerDAO {
    private Log log = LogFactory.getLog(CustomerDAO.class);

    public List findByHql(String hql) {
        log.debug("finding LoginUser instance by hql");
        SessionFactory sf = new Configuration()
                        .configure().buildSessionFactory();
        Session session = sf.openSession();
        try {
            String queryString = hql;
            Query queryObject = session.createQuery(queryString);
            return queryObject.list();
        } catch (RuntimeException re) {
            log.error("find by hql failed", re);
            throw re;
        } finally{
            session.close();
        }
    }
}
```

11. 在 hibernate-prj1 中新建 cn.edu.zjut.service 包,并在其中创建 UserService.java,用于实现登录逻辑,具体代码如下所示。

```java
package cn.edu.zjut.service;
import java.util.List;
import cn.edu.zjut.po.Customer;
import cn.edu.zjut.dao.CustomerDAO;

public class UserService {
    public boolean login(Customer loginUser) {
        String account = loginUser.getAccount();
        String password = loginUser.getPassword();
        String hql = "from Customer as user where account = '"
                        + account + "' and password = '" + password + "'";
        CustomerDAO dao = new CustomerDAO();
        List list = dao.findByHql(hql);
        if(list.isEmpty())
            return false;
        else
            return true;
    }
}
```

实验五 Hibernate 基础应用——基于 Hibernate 框架的用户登录模块

12. 在 hibernate-prj1 中新建 cn.edu.zjut.action 包,并在其中创建 UserAction.java。定义 login()方法,用于调用登录逻辑。具体代码如下所示。

```java
package cn.edu.zjut.action;
import cn.edu.zjut.po.Customer;
import cn.edu.zjut.service.UserService;

public class UserAction {
    private Customer loginUser;

    public Customer getLoginUser() {
        return loginUser;
    }
    public void setLoginUser(Customer loginUser) {
        this.loginUser = loginUser;
    }

    public String login() {
        UserService userServ = new UserService();
        if (userServ.login(loginUser)) {
            return "success";
        }
        return "fail";
    }
}
```

13. 在工程 hibernate-prj1 的 src 目录中创建 struts.xml 文件,用于配置 Action 并设置页面导航,代码片段如下所示。

```xml
<?xml version="1.0" encoding="UTF-8"?>
<!DOCTYPE struts PUBLIC
    "-//Apache Software Foundation//DTD Struts Configuration 2.3//EN"
    "http://struts.apache.org/dtds/struts-2.3.dtd">
<struts>
    <package name="strutsBean" extends="struts-default" namespace="/">
        <action name="login" class="cn.edu.zjut.action.UserAction" method="login">
            <result name="success">/loginSuccess.jsp</result>
            <result name="fail">/login.jsp</result>
        </action>
    </package>
</struts>
```

14. 编辑 Web 应用的 web.xml 文件,增加 Struts2 核心 Filter 的配置,代码片段如下所示。

```xml
<!-- 定义 Struts2 的核心 Filter -->
<filter>
    <filter-name>struts2</filter-name>
    <filter-class>
        org.apache.struts2.dispatcher.ng.filter
            .StrutsPrepareAndExecuteFilter
```

```
            </filter-class>
    </filter>
    <!-- 让 Struts2 的核心 Filter 拦截所有请求 -->
    <filter-mapping>
            <filter-name>struts2</filter-name>
            <url-pattern>/*</url-pattern>
    </filter-mapping>
```

15. 将 hibernate-prj1 部署在 Tomcat 服务器上。
16. 通过浏览器访问 login.jsp 页面，并记录运行结果。

四、实验要求

1. 填写并上交实验报告，报告中应包括如下内容。
（1）运行结果截图。
（2）观察工程 hibernate-prj1 中的 Hibernate 配置文件 hibernate.cfg.xml。查找相关资料，总结配置文件中各元素及其属性的作用，并记录下来。
（3）观察工程 hibernate-prj1 中的 Java 持久化类 Customer.java、Hibernate 映射文件 Customer.hbm.xml，总结持久化类与数据库表的映射关系，以及映射文件中主要元素及其属性的作用，并记录下来。
（4）根据实验过程，总结 Action、Service 和 DAO 之间的调用关系，思考实验一的扩展实验中的 DAO 类与本实验中 DAO 类的区别，并记录下来。
（5）碰到的问题及解决方案或对问题的思考。
（6）实验收获及总结。
2. 上交程序源代码，代码中应有相关注释。

提高实验——持久化对象与 Hibernate 映射文件

一、实验目的

1. 进一步熟悉 Hibernate 应用的基本开发方法。
2. 掌握持久化类与持久化对象的概念，能按照规范进行持久化类的设计开发。
3. 掌握 Hibernate 映射文件的作用，熟悉映射文件中主要元素及其属性的含义和作用，并能进行正确应用。
4. 掌握 Hibernate 中主键的各种生成策略。

二、基本知识与原理

1. 在应用程序中，用来实现业务实体的类被称为持久化类（Persistent Class），如客户信息管理系统中的 Customer 类等。
2. Hibernate 框架中的持久化类与数据库表对应，常用 POJO 编程模式实现。它符合 JavaBean 规范，可提供 public 的无参构造方法，还可提供符合命名规范的 getters 和 setters 方法。
3. 持久化类与数据库表对应，类的属性与表的字段对应。持久化类的对象被称为持久化对象 PO(Persistent Objects)，PO 对应表中的一条记录。

4. 持久化对象映射数据库中的记录，其映射关系依赖于 Hibernate 框架的映射文件配置。映射文件是 XML 文件，往往以 *.hbm.xml 形式命名。其中，* 是持久化对象的类名。

5. Hibernate 映射文件中，元素 <id> 表示持久化类中的主键，<id> 的子元素 <generator> 表示主键的生成策略，其取值可以是 assigned(用户赋值)、increment(自动递增)，等等。

6. 若数据库表中有多个列组成主键，则需要将其对应的持久化类中相应的多个属性封装成一个类，作为复合主键。

三、实验内容及步骤

1. 在 MySQL 的 hibernatedb 数据库中创建一个名称为 item 的数据表，表结构如表 5-3 所示。

表 5-3　item 数据表

字段名称	类型	中文含义
ISBN	VARCHAR(20)，Primary key，Not Null	ISBN 号
title	VARCHAR(30)	书名
description	VARCHAR(100)	说明
cost	FLOAT	单价
image	BLOB	图片

2. 在表 item 中添加 2 条记录，具体内容如表 5-4 所示。

表 5-4　itemr 中的记录

ISBN 号	书　名	说　明	单　价
978-7-121-12345-1	Java EE 技术实验指导教程	Web 程序设计知识回顾、轻量级 Java EE 应用框架、企业级 EJB 组件编程技术、Java EE 综合应用开发	19.95
978-7-121-12345-2	Java EE 技术	Struts 框架、Hibernate 框架、Spring 框架、会话 Bean、实体 Bean、消息驱动 Bean	29.95

3. 在 hibernate-prj1 的 cn.edu.zjut.po 包中手动创建 Java 持久化类 Item.java，使其与 item 数据表相映射，代码片段如下所示。

```
package cn.edu.zjut.po;
import java.sql.Blob;
public class Item {
    private String itemID;
    private String title;
    private String description;
    private float cost;
    private Blob image;
```

```java
        public Item() {
        }
        public Item(String itemID) {
            this.itemID = itemID;
        }
        public Item(String itemID, String title, String description,
                    float cost, Blob image) {
            this.itemID = itemID;
            this.title = title;
            this.description = description;
            this.cost = cost;
            this.image = image;
        }
        //省略 setters/getters 方法
}
```

4. 在 Item.java 的同一目录下,手动创建 Hibernate 映射文件 Item.hbm.xml,并参照 Customer.hbm.xml(实验五基础实验步骤 8)填写 Item.hbm.xml 中的内容。

5. 修改 hibernate-prj1 的 Hibernate 配置文件 hibernate.cfg.xml,增加 Item.hbm.xml 映射文件声明,代码片段如下所示。

```xml
<?xml version = "1.0" encoding = "utf-8"?>
<!DOCTYPE hibernate-configuration PUBLIC
"-//Hibernate/Hibernate Configuration DTD 3.0//EN"
"http://hibernate.sourceforge.net/hibernate-configuration-3.0.dtd">
<hibernate-configuration>
    <session-factory name = "HibernateSessionFactory">
            ⋮
        <mapping resource = "cn/edu/zjut/po/Customer.hbm.xml" />
        <mapping resource = "cn/edu/zjut/po/Item.hbm.xml" />
    </session-factory>
</hibernate-configuration>
```

6. 在 hibernate-prj1 的 cn.edu.zjut.dao 包中创建数据库操作类 ItemDAO.java,具体代码如下所示。

```java
package cn.edu.zjut.dao;

import java.util.List;
import org.hibernate.Query;
import org.hibernate.SessionFactory;
import org.hibernate.Session;
import org.hibernate.cfg.Configuration;
import org.apache.commons.logging.Log;
import org.apache.commons.logging.LogFactory;

public class ItemDAO {
    private static final Log log = LogFactory.getLog(ItemDAO.class);

    public List findAll() {
```

```
        log.debug("finding all Item instances");
        SessionFactory sf = new Configuration()
                            .configure().buildSessionFactory();
        Session session = sf.openSession();
        try {
            String queryString = "from Item";
            Query queryObject = session.createQuery(queryString);
            return queryObject.list();
        } catch (RuntimeException re) {
            log.error("find all failed", re);
            throw re;
        } finally{
            session.close();
        }
    }
}
```

7. 在 hibernate-prj1 的 cn.edu.zjut.service 包中创建 ItemService.java，用于获取所有商品信息，具体代码如下所示。

```
package cn.edu.zjut.service;

import java.util.List;
import java.util.ArrayList;
import cn.edu.zjut.dao.ItemDAO;

public class ItemService {
    private List items = new ArrayList();

    public List getAllItems() {
        ItemDAO dao = new ItemDAO();
        List items = dao.findAll();
        return items;
    }
}
```

8. 在 hibernate-prj1 的 cn.edu.zjut.action 包中创建 ItemAction.java，并在其中定义 getAllItems()方法，用于调用"获取所有商品信息"逻辑。参照 UserAction 类（实验五基础实验步骤 16）填写 ItemAction 类的内容。

9. 在 hibernate-prj1 中新建 itemList.jsp 页面，作为显示所有商品信息的视图，代码片段如下所示。

```
<body>
<center>商品列表</center>
<table border = 1>
<tr>
<th>编号</th>
<th>书名</th>
<th>说明</th>
<th>单价</th>
```

```
</tr>
<s:iterator value = "items">
<tr>
    <td><s:property value = "itemID"/></td>
    <td><s:property value = "title"/></td>
    <td><s:property value = "description"/></td>
    <td><s:property value = "cost"/></td>
</tr>
</s:iterator>
</table>
</body>
```

10. 修改 hibernate-prj1 的 loginSuccess.jsp 页面,在视图中增加超链接,用于查看所有商品信息,代码片段如下所示。

```
<a href = "./allItems">查看所有商品信息</a>
```

11. 修改 hibernate-prj1 的 struts.xml 文件,增加 ItemAction 的配置并设置页面导航。

12. 将 hibernate-prj1 重新部署在 Tomcat 服务器上。

13. 通过浏览器访问 login.jsp 页面,并记录运行结果。

14. 假设使用持久化类 Item.java 中的 itemID 属性和 title 属性作为复合主键,则需要将这两个属性封装成一个类,代码片段如下所示。

```
package cn.edu.zjut.po;
public class ItemPK{
    private String itemID;
    private String title;
    //省略 setters/getters 方法
}
```

15. 修改 Item.java,将 ItemPK 主键作为其属性之一,代码片段如下所示。

```
package cn.edu.zjut.po;
⋮
public class Item {
    private ItemPK ipk;
    private String description;
    private float cost;
    private Blob image;
    ⋮
}
```

16. 修改 Hibernate 映射文件 Item.hbm.xml,将主键类中每个属性和表中的列对应,并指定复合主键的类型,代码片段如下所示。

```
<hibernate-mapping>
    <class name = "cn.edu.zjut.po.Item" table = "item">
        <composite-id name = "ipk" class = "cn.edu.zjut.po.ItemPK">
            <key-property name = "itemID" column = "ISBN"/>
            <key-property name = "title" column = "title"/>
        </composite-id>
```

```
            ⋮
        </class>
</hibernate-mapping>
```

17. 将 hibernate-prj1 重新部署在 Tomcat 服务器上。通过浏览器访问 login.jsp 页面，并记录运行结果。

四、实验要求

1. 填写并上交实验报告，报告中应包括如下内容。
（1）运行结果截图。
（2）结合实验过程，查找相关资料，总结 POJO 模式下持久化类的规范，并记录下来。
（3）结合实验过程，查找相关资料，总结映射文件中主要元素（如 class、id、generator、property）及其属性的含义与作用，并记录下来。
（4）结合实验过程，查找相关资料，总结设置复合主键的方法和步骤，并记录下来。
（5）查找相关资料，总结 Hibernate 映射文件中主键的各种生成策略的作用，并记录下来。
（6）碰到的问题及解决方案或对问题的思考。
（7）实验收获及总结。
2. 上交程序源代码，代码中应有相关注释。

扩展实验——粒度设计

一、实验目的

1. 进一步熟悉持久化类与 Hibernate 映射文件的开发方法。
2. 学习在实际应用中进行粒度细分的方法，将一张表映射到多个类。

二、基本知识与原理

1. 在实际应用中，并不总是一张表与一个实体类映射，往往可能会出现一张表跟多个实体类映射的情况，此称为粒度设计。
2. 如果将表中的某些字段联合起来能表示持久化类中的某一个属性，那么就可以进行基于设计的粒度设计，即将表跟多个类映射，类和类之间使用关联关系。此时只需要一个映射文件，其中使用 component 元素进行映射。
3. 如果表中的某些字段不经常使用，而且占有的空间较大，则可以使用基于性能的粒度设计，即一个表可以映射为多个类，每个类对应一个 *.hbm.xml 文件。可以根据实际情况，使用不同的类。

三、实验内容及步骤

1. 在 hibernatedb 数据库里的 customer 数据表中，包括 phone、email、address、zipcode、fax 在内的字段都属于用户的联系方式。基于设计的粒度设计将用户的联系方式单独封装到类 ContactInfo 中，代码片段如下所示。

```java
package cn.edu.zjut.po;
public class ContactInfo {

    private String phone;
    private String email;
    private String address;
    private String zipcode;
    private String fax;

    public ContactInfo() {
        super();
    }

    public ContactInfo(String phone, String email, String address,
            String zipcode, String fax) {
        super();
        this.phone = phone;
        this.email = email;
        this.address = address;
        this.zipcode = zipcode;
        this.fax = fax;
    }
    //省略 setters/getters 方法
}
```

2. 修改 Customer.java，将 ContactInfo 实例作为 Customer 的属性，用于表示用户的联系方式，代码片段如下所示。

```java
package cn.edu.zjut.po;
public class Customer {

    private int customerId;
    private String account;
    private String password;
    private String name;
    private Boolean sex;
    private Date birthday;
    private ContactInfo contactInfo;

    //省略构造函数和 setters/getters 方法
}
```

3. 修改 Hibernate 映射文件 Customer.hbm.xml，将 customer 表与两个类（Customer 和 ContactInfo）映射，具体代码如下所示。

```xml
<?xml version="1.0"?>
<!DOCTYPE hibernate-mapping PUBLIC "-//Hibernate/Hibernate Mapping DTD 3.0//EN"
"http://hibernate.sourceforge.net/hibernate-mapping-3.0.dtd">
<hibernate-mapping>
    <class name="cn.edu.zjut.po.Customer" table="customer" catalog="hibernatedb">
        <id name="customerId" type="int">
```

```xml
            <column name="customerID" />
            <generator class="increment" />
        </id>
        <property name="account" type="string">
            <column name="account" length="20" unique="true" />
        </property>
        <property name="password" type="string">
            <column name="password" length="20" />
        </property>
        <property name="name" type="string">
            <column name="name" length="20" />
        </property>
        <property name="sex" type="java.lang.Boolean">
            <column name="sex" />
        </property>
        <property name="birthday" type="date">
            <column name="birthday" length="10" />
        </property>
        <component name="contactInfo" class="cn.edu.zjut.po.ContactInfo">
            <property name="phone" type="string">
                <column name="phone" length="20" />
            </property>
            <property name="email" type="string">
                <column name="email" length="100" />
            </property>
            <property name="address" type="string">
                <column name="address" length="200" />
            </property>
            <property name="zipcode" type="string">
                <column name="zipcode" length="10" />
            </property>
            <property name="fax" type="string">
                <column name="fax" length="20" />
            </property>
        </component>
    </class>
</hibernate-mapping>
```

4. 修改 hibernate-prj1 的 Hibernate 配置文件 hibernate.cfg.xml,开启增、删、改操作的事务自动提交功能,代码片段如下所示。

```xml
<?xml version="1.0" encoding="utf-8"?>
<!DOCTYPE hibernate-configuration PUBLIC
"-//Hibernate/Hibernate Configuration DTD 3.0//EN"
"http://hibernate.sourceforge.net/hibernate-configuration-3.0.dtd">
<hibernate-configuration>
    <session-factory name="HibernateSessionFactory">
        ⋮
        <property name="connection.autocommit">true</property>
    </session-factory>
</hibernate-configuration>
```

5. 在 hibernate-prj1 中新建 register.jsp 页面,作为用户注册的视图(可参考实验二中提高实验里的页面代码,并根据数据库表 customer 设计页面内容);新建 regSuccess.jsp 页面,作为注册成功的视图(可重用实验二中提高实验里的页面代码)。

6. 修改 cn.edu.zjut.dao 包中的 CustomerDAO.java,增加"添加新用户"的操作,代码片段如下所示。

```java
public class CustomerDAO {
    ⋮
    public void save(Customer customer) {
        log.debug("saving customer instance");
        SessionFactory sf = new Configuration().
                                configure().buildSessionFactory();
        Session session = sf.openSession();
        try {
            session.save(customer);
            session.flush();
            log.debug("save successful");
        } catch (RuntimeException re) {
            log.error("save failed", re);
            throw re;
        } finally{
            session.close();
        }
    }
}
```

7. 修改 cn.edu.zjut.service 包中的 UserService.java,增加用户注册逻辑,代码片段如下所示。

```java
public class UserService {
    ⋮
    public void register(Customer loginUser) {
        CustomerDAO dao = new CustomerDAO();
        dao.save(loginUser);
    }
}
```

8. 修改 cn.edu.zjut.action 包中的 UserAction.java,定义 register()方法,用于调用用户注册逻辑,代码片段如下所示。

```java
public class UserAction {
    ⋮
    public String register() {
        UserService userServ = new UserService();
        userServ.register(loginUser);
        return "registersuccess";
    }
}
```

9. 修改工程 hibernate-prj1 中的 struts.xml 文件,为"用户注册"增加 Action 配置并设

10. 将 hibernate-prj1 重新部署在 Tomcat 服务器上。通过浏览器访问 register.jsp 页面，记录运行结果，并在数据库里的 customer 表中检查是否写入了新的记录。

11. 在 hibernatedb 数据库里的 item 数据表中，image 字段表示商品的照片。它使用 Blob 类型，所占空间较大，如果该字段不经常被使用，则基于性能的粒度设计，将表 item 映射为 2 个类。其中一个类为 ItemDetail，映射表中所有的字段；而另一个类 ItemBasic，映射表中除了 image 之外的字段，具体代码如下所示。

```java
//ItemBasic
package cn.edu.zjut.po;
public class ItemBasic {
    private String itemID;
    private String title;
    private String description;
    private float cost;

    public ItemBasic() {
    }
    public ItemBasic(String itemID) {
        this.itemID = itemID;
    }
    public ItemBasic(String itemID, String title, String description, float cost) {
        this.itemID = itemID;
        this.title = title;
        this.description = description;
        this.cost = cost;
    }
    //省略 setters/getters 方法
}
```

```java
//ItemDetail
package cn.edu.zjut.po;
Import java.sql.Blob;
public class ItemDetail extends ItemBasic{
    private Blob image;

    public ItemDetail() {
    }
    public ItemDetail(String itemID, String title, String description,
                float cost, Blob image) {
        super(itemID, title, description, cost);
        this.image = image;
    }
    //省略 setters/getters 方法
}
```

12. 在 ItemBasic.java 和 ItemDetail.java 的同一目录下，手动创建 Hibernate 映射文件 ItemBasic.hbm.xml 和 ItemDetail.hbm.xml。ItemDetail.hbm.xml 的具体代码如下所示

（ItemBasic.hbm.xml 略）。

```xml
<?xml version="1.0"?>
<!DOCTYPE hibernate-mapping PUBLIC "-//Hibernate/Hibernate Mapping DTD 3.0//EN"
"http://hibernate.sourceforge.net/hibernate-mapping-3.0.dtd">
<hibernate-mapping>
    <class name="cn.edu.zjut.po.ItemDetail" table="item" catalog="hibernatedb">
        <id name="itemID" type="string">
            <column name="ISBN" length="20" />
            <generator class="assigned" />
        </id>
        <property name="title" type="string">
            <column name="title" length="30" />
        </property>
        <property name="description" type="string">
            <column name="description" length="100" />
        </property>
        <property name="cost" type="float">
            <column name="cost" />
        </property>
        <property name="image" type="java.sql.Blob">
            <column name="image" />
        </property>
    </class>
</hibernate-mapping>
```

13. 修改 hibernate-prj1 的 Hibernate 配置文件 hibernate.cfg.xml，删除 Item.hbm.xml 映射文件声明，增加 ItemDetail.hbm.xml 和 ItemBasic.hbm.xml 映射文件声明，代码片段如下所示。

```xml
<?xml version="1.0" encoding="utf-8"?>
<!DOCTYPE hibernate-configuration PUBLIC
"-//Hibernate/Hibernate Configuration DTD 3.0//EN"
"http://hibernate.sourceforge.net/hibernate-configuration-3.0.dtd">
<hibernate-configuration>
    <session-factory name="HibernateSessionFactory">
        ⋮
        <mapping resource="cn/edu/zjut/po/Customer.hbm.xml" />
        <mapping resource="cn/edu/zjut/po/ItemDetail.hbm.xml" />
        <mapping resource="cn/edu/zjut/po/ItemBasic.hbm.xml" />
    </session-factory>
</hibernate-configuration>
```

14. 将 hibernate-prj1 重新部署在 Tomcat 服务器上。
15. 通过浏览器访问 login.jsp 页面，并记录运行结果。

四、实验要求

1. 填写并上交实验报告，报告中应包括如下内容。
（1）运行结果截图。

(2) 结合实验过程,总结两种粒度设计的方法及特点。

(3) 根据实验步骤 4,查找相关资料,写出 Hibernate 配置文件 hibernate.cfg.xml 中的 connection.autocommit 属性的作用。

(4) 碰到的问题及解决方案或对问题的思考。

(5) 实验收获及总结。

2. 上交程序源代码,代码中应有相关注释。

实验六

Hibernate 的体系结构——登录用户信息的增、删、改、查

基础实验——Hibernate 常用 API

一、实验目的

1. 进一步掌握 Hibernate 应用的开发方法,理解 Hibernate 配置文件中主要元素的作用,会开发持久化类,并进行相应的 Hibernate 映射文件配置。

2. 学习并掌握 Hibernate 框架的常用 API,掌握利用 Hibernate 基本 API 载入配置文件、建立数据库连接的基本步骤。

3. 理解 Hibernate 基本 API 中 Session 的主要作用,掌握利用 Session 进行数据库操作的基本步骤。

二、基本知识与原理

1. Hibernate 进行持久化操作,通常有如下操作步骤。
(1) 开发持久化类和映射文件。
(2) 获取 Configuration。
(3) 获取 SessionFactory。
(4) 获取 Session,打开事务。
(5) 用面向对象的方式操作数据库。
(6) 关闭事务,关闭 Session。

2. 对 PO 的操作必须在 Session 管理下才能同步到数据库。Session 是应用程序与持久储存层之间交互操作的一个单线程对象,它底层封装了 JDBC 连接,主要用来对 PO 进行创建、读取、删除等操作。

3. Session 由 SessionFactory 生成。SessionFactory 是数据库编译后的内存镜像。通常,一个应用对应一个 SessionFactory 对象。

4. SessionFactory 对象由 Configuration 对象生成。Configuration 对象负责加载 Hibernate 配置文件。每个 Hibernate 配置文件对应一个 Configuration 对象。

三、实验内容及步骤

1. 在 MySQL 中创建一个名称为 hibernatedb 的数据库,并在该数据库中创建一个名称为 customer 的数据表,表结构如表 6-1 所示。

表 6-1　customer 数据表

字段名称	类　　型	中文含义
customerID	INTEGER(11),Primary key,Not Null	用户编号
account	VARCHAR(20)	登录用户名
password	VARCHAR(20)	登录密码
name	VARCHAR(20)	真实姓名
sex	BOOLEAN(1)	性别
birthday	DATE	出生日期
phone	VARCHAR(20)	联系电话
email	VARCHAR(100)	电子邮箱
address	VARCHAR(200)	联系地址
zipcode	VARCHAR(10)	邮政编码
fax	VARCHAR(20)	传真号码

2. 在 Eclipse 中新建 Web 工程 hibernate-prj2,并添加 MySQL 驱动程序库文件、commons-logging-1.2.jar、Struts2 核心包和 Hibernate 核心包到工程中。

3. 在 hibernate-prj2/src 中新建配置文件 hibernate.cfg.xml,具体代码可参照实验五中基础实验里的 hibernate.cfg.xml。

4. 在 hibernate-prj2 中新建 cn.edu.zjut.po 包,并在其中创建持久化类 Customer.java 以及 Hibernate 映射文件 Customer.hbm.xml,具体代码可参照实验五中基础实验里的内容。

5. 修改配置文件 hibernate.cfg.xml,增加 Customer.hbm.xml 映射文件的声明。

6. 在 hibernate-prj2 中新建 cn.edu.zjut.dao 包,并在其中创建 DAO 操作辅助类 HibernateUtil.java,用于生成 Configuration 对象和 SessionFactory 对象,具体代码如下所示。

```java
package cn.edu.zjut.dao;
import org.hibernate.HibernateException;
import org.hibernate.Session;
import org.hibernate.cfg.Configuration;

public class HibernateUtil {
    private static String CONFIG_FILE_LOCATION = "/hibernate.cfg.xml";
    private static final ThreadLocal<Session>
                threadLocal = new ThreadLocal<Session>();
    private static Configuration configuration = new Configuration();
    private static org.hibernate.SessionFactory sessionFactory;
    private static String configFile = CONFIG_FILE_LOCATION;

    static {
        try {
            configuration.configure(configFile);
            sessionFactory = configuration.buildSessionFactory();
        } catch (Exception e) {
```

```java
            System.err
                    .println("%%%% Error Creating SessionFactory %%%%");
            e.printStackTrace();
        }
    }
    private HibernateUtil() { }

    public static org.hibernate.SessionFactory getSessionFactory() {
        return sessionFactory;
    }
    public static void rebuildSessionFactory() {
        try {
            configuration.configure(configFile);
            sessionFactory = configuration.buildSessionFactory();
        } catch (Exception e) {
            System.err
                    .println("%%%% Error Creating SessionFactory %%%%");
            e.printStackTrace();
        }
    }

    public static void setConfigFile(String configFile) {
        HibernateUtil.configFile = configFile;
        sessionFactory = null;
    }
    public static Configuration getConfiguration() {
        return configuration;
    }
}
```

7. 完善辅助类 HibernateUtil.java，在其中添加代码，用于获取 Session 对象和关闭 Session 对象，代码片段如下所示。

```java
package cn.edu.zjut.dao;
public class HibernateUtil {
    ⋮
    public static Session getSession() throws HibernateException {
        Session session = (Session) threadLocal.get();
        if (session == null || !session.isOpen()) {
            if (sessionFactory == null) {
                rebuildSessionFactory();
            }
            session = (sessionFactory != null)
                            ? sessionFactory.openSession() : null;
            threadLocal.set(session);
        }
        return session;
    }
    public static void closeSession() throws HibernateException {
        Session session = (Session) threadLocal.get();
        threadLocal.set(null);
```

```
            if (session != null) {
                session.close();
            }
        }
    }
}
```

8. 在 hibernate-prj2 的 cn.edu.zjut.dao 包中创建 BaseHibernateDAO.java,作为数据库操作基础类,具体代码如下所示。

```
package cn.edu.zjut.dao;
import org.hibernate.Session;
public class BaseHibernateDAO{
    public Session getSession() {
        return HibernateUtil.getSession();
    }
}
```

9. 在 hibernate-prj2 的 cn.edu.zjut.dao 包中创建 CustomerDAO.java,继承数据库操作基础类 BaseHibernateDAO.java,实现 Customer 增、删、改、查操作,代码片段如下所示。

```
package cn.edu.zjut.dao;
import java.util.List;
import org.hibernate.Query;
import org.apache.commons.logging.Log;
import org.apache.commons.logging.LogFactory;
import cn.edu.zjut.po.Customer;
public class CustomerDAO extends BaseHibernateDAO{
    private Log log = LogFactory.getLog(CustomerDAO.class);

    public List findByHql(String hql) {
        log.debug("finding Customer instance by hql");
        try {
            String queryString = hql;
            Query queryObject = getSession().createQuery(queryString);
            return queryObject.list();
        } catch (RuntimeException re) {
            log.error("find by hql failed", re);
            throw re;
        }
    }
    public void save(Customer instance) {
        log.debug("saving Customer instance");
        try {
            getSession().save(instance);
            log.debug("save successful");
        } catch (RuntimeException re) {
            log.error("save failed", re);
            throw re;
        }
    }
    public void update(Customer instance) { //省略代码 }
```

```java
        public void delete(Customer instance) { //省略代码 }
}
```

10. 在 hibernate-prj2 中新建 cn.edu.zjut.service 包,并在其中创建 UserService.java,用于实现登录、注册、个人信息修改和删除逻辑,代码片段如下所示。

```java
package cn.edu.zjut.service;
import com.opensymphony.xwork2.ActionContext;
   ⋮
public class UserService {
    private Map<String, Object> request, session;

    public boolean login(Customer loginUser) {
        ActionContext ctx = ActionContext.getContext();
        session = (Map) ctx.getSession();
        request = (Map) ctx.get("request");
        String account = loginUser.getAccount();
        String password = loginUser.getPassword();
        String hql = "from Customer as user where account = '" + account
                + "' and password = '" + password + "'";
        CustomerDAO dao = new CustomerDAO();
        List list = dao.findByHql(hql);
        dao.getSession().close();
        if (list.isEmpty())
            return false;
        else{
            session.put("user", account);
            request.put("tip", "登录成功!");
            loginUser = (Customer)list.get(0);
            request.put("loginUser", loginUser);
            return true;
        }
    }
    public boolean register(Customer loginUser) { //省略代码 }
    public boolean update(Customer loginUser) { //省略代码 }

    public boolean delete(Customer loginUser) {
        ActionContext ctx = ActionContext.getContext();
        session = (Map) ctx.getSession();
        request = (Map) ctx.get("request");
        CustomerDAO dao = new CustomerDAO();
        Transaction tran = null;
        try {
            tran = dao.getSession().beginTransaction();
            dao.delete(loginUser);
            tran.commit();
            session.remove("user");
            request.put("tip", "删除个人信息成功,请重新登录!");
            return true;
        } catch (Exception e) {
            if(tran != null) tran.rollback();
```

```
            return false;
        } finally {
            dao.getSession().close();
        }
    }
}
```

11. 在 hibernate-prj2 中新建 cn.edu.zjut.action 包,并在其中创建 UserAction.java,代码片段如下所示。

```
package cn.edu.zjut.action;
⋮
public class UserAction {
    private Customer loginUser;
    //省略 getters/setters 方法

    public String login() {
        UserService userServ = new UserService();
        if (userServ.login(loginUser))
            return "loginsuccess";
        else
            return "loginfail";
    }
    public String register() { //省略代码 }
    public String update() { //省略代码 }
    public String delete() { //省略代码 }
}
```

12. 在 hibernate-prj2 中新建 login.jsp 页面,作为用户登录视图,代码片段如下所示。

```
<body>
<s:property value="#request.tip"/><p>
<s:form action="Userlogin" method="post">
    <s:textfield name="loginUser.account" label="请输入用户名"/>
    <s:password name="loginUser.password" label="请输入密码"/>
    <s:submit value="登录"/>
</s:form>
</body>
```

13. 在 hibernate-prj2 中新建 loginSuccess.jsp 页面,作为登录成功视图,并要求能在该页面中修改个人信息或删除个人信息,代码片段如下所示。

```
<body>
<s:property value="#request.tip"/><p>
修改个人信息<p>
<s:form action="Userupdate" method="post">
    <s:hidden name="loginUser.customerId"
        value="%{#request.loginUser.customerId}"/>
    <s:textfield name="loginUser.account" label="用户名不能修改"
        value="%{#request.loginUser.account}" readonly="true"/>
    <s:textfield type="password" name="loginUser.password"
        label="修改密码" value="%{#request.loginUser.password}"/>
```

```
        <!-- 省略其他表单域 -->
        <s:submit value="修改"/>
    </s:form>

    <s:form action="Userdelete" method="post">
        <s:hidden name="loginUser.customerId"
            value="%{#request.loginUser.customerId}"/>
        <s:submit value="删除"/>
    </s:form>
</body>
```

14. 在 hibernate-prj2 中新建 register.jsp 页面,作为用户注册视图(可参考实验二中提高实验里的页面代码,并根据数据库表 customer 设计页面内容)。

15. 在 hibernate-prj2 中新建 registerSuccess.jsp 页面,作为注册成功的视图(代码略)。

16. 在 hibernate-prj2 中新建 CURDFail.jsp 页面,作为修改个人信息和删除个人信息操作失败的视图(代码略)。

17. 在 hibernate-prj2 的 src 目录中创建 struts.xml 文件,用于配置 Action 并设置页面导航,使得:登录成功时转向 loginSuccess.jsp 页面,失败时转向 login.jsp 页面重新登录;注册成功时转向 registerSuccess.jsp 页面,失败时转向 register.jsp 页面重新注册;修改个人信息成功时转向 loginSuccess.jsp 页面,失败时转向 CURDFail.jsp 页面;删除个人信息成功时转向 login.jsp 页面重新登录,失败时转向 CURDFail.jsp 页面。

18. 编辑 Web 应用的 web.xml 文件,增加 Struts2 核心 Filter 的配置。

19. 将 hibernate-prj2 部署在 Tomcat 服务器上。

20. 通过浏览器访问 register.jsp 页面,注册一个新用户(假设用户名各不相同),然后用该用户名登录,并修改用户信息或删除用户。查看并记录运行结果。

四、实验要求

1. 填写并上交实验报告,报告中应包括如下内容。
(1) 运行结果截图。
(2) 结合实验过程,查找相关资料,总结利用 Hibernate 基本 API 载入配置文件、建立数据库连接并进行数据库操作的基本步骤。
(3) 结合实验过程,查找相关资料,总结 Session 中的主要方法及其与数据库操作的对应关系。
(4) 根据实验步骤 8~9,总结 Session 和 DAO 之间的调用关系,思考事务操作相关代码所处的位置,并记录下来。
(5) 根据实验步骤 12,思考表单域中 value 值的写法,总结显示 value 值的实现方法并记录下来。
(6) 碰到的问题及解决方案或对问题的思考。
(7) 实验收获及总结。

2. 上交程序源代码,代码中应有相关注释。

提高实验——HQL 语言

一、实验目的

1. 学习 HQL 的使用方法，掌握 select 子句、from 子句、where 子句、order by 子句、聚合函数、group by 子句、子查询等基本语法，并能正确运用。

2. 理解 HQL 语言是一种面向对象的查询语言，能正确区分 HQL 语言与 SQL 语言的差别。

二、基本知识与原理

1. HQL(Hibernate Query Language)语言是 Hibernate 框架定义的查询语言。

2. HQL 语言的语法结构与 SQL 语言非常类似，但 HQL 是面向对象的查询语言。HQL 语句中使用的是 Java 类名和属性名，其对大小写敏感。

3. HQL 语言包括 select 子句、from 子句、where 子句、order by 子句、聚合函数、group by 子句、子查询、连接查询等。

三、实验内容及步骤

1. 在 MySQL 的 hibernatedb 数据库中创建一个名称为 item 的数据表，表结构如表 6-2 所示。

表 6-2 item 数据表

字段名称	类 型	中文含义
ISBN	VARCHAR(20)，Primary key，Not Null	ISBN 号
title	VARCHAR(30)	书名
description	VARCHAR(100)	说明
cost	FLOAT	单价
image	BLOB	图片

2. 在表 item 添加 2 条记录，具体内容如表 6-3 所示。

表 6-3 itemr 中的记录

ISBN 号	书 名	说 明	单 价
978-7-121-12345-1	Java EE 技术实验指导教程	Web 程序设计知识回顾、轻量级 Java EE 应用框架、企业级 EJB 组件编程技术、Java EE 综合应用开发	19.95
978-7-121-12345-2	Java EE 技术	Struts 框架、Hibernate 框架、Spring 框架、会话 Bean、实体 Bean、消息驱动 Bean	29.95

3. 在 hibernate-prj2 的 cn.edu.zjut.po 包中创建持久化类 Item.java 以及 Hibernate 映射文件 Item.hbm.xml，具体代码可参照实验五中提高实验里的内容。

4. 修改 hibernate-prj2 的 Hibernate 配置文件 hibernate.cfg.xml，增加 Item.hbm.xml 映射文件的声明。

5. 在 hibernate-prj2 的 cn.edu.zjut.dao 包中创建数据库操作类 ItemDAO.java，继承数据库操作基础类 BaseHibernateDAO.java，具体代码如下所示。

```java
package cn.edu.zjut.dao;
import java.util.List;
import org.hibernate.Query;
import org.apache.commons.logging.Log;
import org.apache.commons.logging.LogFactory;

public class ItemDAO extends BaseHibernateDAO{
    private static final Log log = LogFactory.getLog(ItemDAO.class);

    public List findByHql(String hql) {
        log.debug("finding Item instance by hql");
        try {
            String queryString = hql;
            Query queryObject = getSession().createQuery(queryString);
            return queryObject.list();
        } catch (RuntimeException re) {
            log.error("find by hql failed", re);
            throw re;
        }
    }
}
```

6. 在 hibernate-prj2 的 cn.edu.zjut.service 包中创建 ItemService.java。使用 from 子句实现最简单的查询，获取所有商品信息，具体代码如下所示。

```java
package cn.edu.zjut.service;
import java.util.List;
import java.util.ArrayList;
import cn.edu.zjut.dao.ItemDAO;

public class ItemService {
    private List items = new ArrayList();

    public List findByHql() {
        ItemDAO dao = new ItemDAO();
        String hql = "from cn.edu.zjut.po.Item";
        List list = dao.findByHql(hql);
        dao.getSession().close();
        return list;
    }
}
```

7. 在 hibernate-prj2 的 cn.edu.zjut.action 包中创建 ItemAction.java，并在其中定义

findItems()方法,用于调用获取商品信息逻辑,代码片段如下所示。

```
package cn.edu.zjut.action;
import java.util.List;
import cn.edu.zjut.service.ItemService;

public class ItemAction{
    private List items;
    //省略 getters/setters 方法
    public String findItems() {
        ItemService itemServ = new ItemService();
        items = itemServ.findByHql();
        System.out.println("Item Action executed!");
        return "success";
    }
}
```

8. 在 hibernate-prj2 中新建 itemList.jsp 页面,作为显示商品信息的视图,代码片段如下所示。

```
<body>
<center>商品列表</center>
<table border=1>
<tr>
<th>编号</th>
<th>书名</th>
<th>说明</th>
<th>单价</th>
</tr>
<s:iterator value="items">
<tr>
    <td><s:property value="itemID"/></td>
    <td><s:property value="title"/></td>
    <td><s:property value="description"/></td>
    <td><s:property value="cost"/></td>
</tr>
</s:iterator>
</table>
</body>
```

9. 修改 hibernate-prj2 的 loginSuccess.jsp 页面,在视图中增加超链接,用于查看商品信息,代码片段如下所示。

```
<a href="./findItems">查看商品信息</a>
```

10. 修改 hibernate-prj2 的 struts.xml 文件,增加 ItemAction 的配置并设置页面导航。

11. 将 hibernate-prj2 重新部署在 Tomcat 服务器上。通过浏览器访问 login.jsp 页面,并记录运行结果。

12. 修改 ItemService.java。将 hql 语句替换成"from Item",省略包名,直接通过类名查询,代码片段如下所示。

```
String hql = "from Item";
```

13. 将 hibernate-prj2 重新部署在 Tomcat 服务器上。通过浏览器访问 login.jsp 页面，并记录运行结果。

14. 修改 ItemService.java 中的 hql 语句。使用"as"为类取名，以便在其他地方使用，代码片段如下所示。

```
String hql = "from Item as item";
```

15. 将 hibernate-prj2 重新部署在 Tomcat 服务器上。通过浏览器访问 login.jsp 页面，并记录运行结果。

16. 修改 ItemService.java 中的 hql 语句。使用 select 子句查询商品名称。代码片段如下所示。

```
String hql = "select item.title from Item as item";
```

17. 由于 select 子句只返回 title 属性，则返回结果将直接封装到该元素类型的集合中，即封装到 List<String>集合中，因此需要修改 itemList.jsp 页面，以显示商品名称，代码片段如下所示。

```
<table border=1>
<tr><th>书名</th></tr>
<s:iterator value="items" id="object">
<tr>
    <td><s:property value="object"/></td>
</tr>
</s:iterator>
</table>
```

18. 将 hibernate-prj2 重新部署在 Tomcat 服务器上。通过浏览器访问 login.jsp 页面，并记录运行结果。

19. 修改 ItemService.java 中的 hql 语句。使用 select 子句查询商品名称和商品价格。代码片段如下所示。

```
String hql = "select item.title, item.cost from Item as item";
```

20. 若 select 子句返回多个属性，则返回结果将被封装到 List<Object[]>集合中，因此需要修改 itemList.jsp 页面，以显示商品名称和商品价格，代码片段如下所示。

```
<table border=1>
<tr><th>书名</th><th>单价</th></tr>
<s:iterator value="items" id="object">
<tr>
    <td><s:property value="#object[0]"/></td>
    <td><s:property value="#object[1]"/></td>
</tr>
</s:iterator>
</table>
```

21. 将 hibernate-prj2 重新部署在 Tomcat 服务器上。通过浏览器访问 login.jsp 页面，

并记录运行结果。

22. 修改 ItemService.java 中的 hql 语句，尝试使用聚集函数、where 子句、order by 子句和子查询进行数据查询，并通过 itemList.jsp 页面显示查询结果。

四、实验要求

1. 填写并上交实验报告，报告中应包括如下内容。
（1）运行结果截图。
（2）查找相关资料，完成实验步骤 22，并记录运行结果。
（3）结合实验过程，查找相关资料，总结 HQL 的常用语句及语法规则。
（4）碰到的问题及解决方案或对问题的思考。
（5）实验收获及总结。
2. 上交程序源代码，代码中应有相关注释。

扩展实验——深入 Hibernate 配置文件

一、实验目的

1. 深入学习 Hibernate 配置文件中的配置属性，了解数据库连接池配置等可选配置属性的作用。
2. 深入学习 Configuration 对象，能通过编程方式创建 Configuration 实例并为该对象设置一系列属性。

二、基本知识与原理

1. Hibernate 配置文件中包括 JDBC 连接属性、数据库方言、Hibernate 事务属性、二级缓存等配置属性。
2. 一个 Configuration 实例代表了一个应用程序中 Java 类型到 SQL 数据库映射的完整集合。
3. 每个 Hibernate 配置文件对应一个 Configuration 对象，但在没有任何配置文件的情况下，也可以通过编程方式创建 Configuration 对象。
4. Configuration 对象提供了若干方法，如下所示。
（1）Configuration addResource(String resourceName)：用于为 Configuration 对象添加一个映射文件。
（2）Configuration setProperty(String property, String value)：用于为 Configuration 对象设置属性。

三、实验内容及步骤

1. 修改 hibernate-prj2 的 Hibernate 配置文件 hibernate.cfg.xml，增加配置属性，使得能在控制台输出 Hibernate 生成的格式良好的 SQL 语句，并在其中添加有助于调试的注释，代码片段如下所示。

```xml
<hibernate-configuration>
    <session-factory name="HibernateSessionFactory">
        ⋮
        <property name="hibernate.show_sql">true</property>
        <property name="hibernate.format_sql">true</property>
        <property name="hibernate.use_sql_comments">true</property>
    </session-factory>
</hibernate-configuration>
```

2. 将 hibernate-prj2 重新部署在 Tomcat 服务器上。通过浏览器访问 login.jsp 页面，查看控制台输出的 SQL 语句。

3. 修改 hibernate.cfg.xml，增加 C3P0 连接池的配置属性，代码片段如下所示。

```xml
<hibernate-configuration>
    <session-factory name="HibernateSessionFactory">
        ⋮
        <property name="hibernate.c3p0.max_size">20</property>
        <property name="hibernate.c3p0.min_size">1</property>
        <property name="hibernate.c3p0.timeout">1800</property>
        <property name="hibernate.c3p0.max_statements">50</property>
    </session-factory>
</hibernate-configuration>
```

4. 将 hibernate-prj2 重新部署在 Tomcat 服务器上，通过浏览器访问 login.jsp 页面，查看控制台的相关输出(INFO)，并记录下来。

5. 删除 hibernate-prj2 的 Hibernate 配置文件 hibernate.cfg.xml。

6. 修改 cn.edu.zjut.dao 包中的 HibernateUtil.java，通过编程方式创建 Configuration 对象，代码片段如下所示。

```java
public class HibernateUtil {
    private static Configuration configuration = new Configuration();
    ⋮
    static {
        try {
            configuration
                //通过 setProperty 方法设置 Hibernate 的连接属性
                .setProperty("hibernate.connection.driver_class",
                        "com.mysql.jdbc.Driver")
                .setProperty("hibernate.connection.url",
"jdbc:mysql://localhost:3306/hibernatedb")
                .setProperty("hibernate.connection.username", "root")
                .setProperty("hibernate.connection.password", "")
                .setProperty("hibernate.dialect",
                        "org.hibernate.dialect.MySQLDialect")
                //通过 addResource 方法添加映射文件
                .addResource("cn/edu/zjut/po/Customer.hbm.xml")
                .addResource("cn/edu/zjut/po/Item.hbm.xml");
            sessionFactory = configuration.buildSessionFactory();
        } catch (Exception e) { ⋯ }
    }
}
```

}

7. 将 hibernate-prj2 重新部署在 Tomcat 服务器上。通过浏览器访问 login.jsp 页面，并记录运行结果。

8. 在 MySQL 的 hibernatedb 数据库中删除 customer 数据表。

9. 修改 HibernateUtil.java。通过编程方式添加 hibernate.hbm2ddl.auto 配置属性，使得能根据映射文件自动建立数据库表，代码片段如下所示。

```
public class HibernateUtil {
    private static Configuration configuration = new Configuration();
      ⋮
    static {
        try {
          ⋮
            configuration
              .setProperty("hibernate.hbm2ddl.auto", "create-drop");
        } catch (Exception e) { ⋯ }
    }
}
```

10. 将 hibernate-prj2 重新部署在 Tomcat 服务器上。通过浏览器访问 register.jsp 页面，并记录运行结果。

四、实验要求

1. 填写并上交实验报告，报告中应包括如下内容。

(1) 运行结果截图。

(2) 结合实验过程，查找相关资料，写出实验步骤 1、3 中所使用的配置属性的作用，并记录下来。

(3) 根据实验步骤 9，查找相关资料，写出"hibernate.hbm2ddl.auto"配置属性及其取值的作用。

(4) 根据实验步骤 6，查找相关资料，总结通过编程方式创建 Configuration 实例或设置配置属性的基本步骤，思考在实际应用中如何将 Hibernate 配置文件设置配置属性与编程方式设置配置属性相结合，并将其记录下来。

(5) 碰到的问题及解决方案或对问题的思考。

(6) 实验收获及总结。

2. 上交程序源代码，代码中应有相关注释。

实验七
Hibernate 关联关系映射
——登录用户的地址管理

基础实验——一对多/多对一关联

一、实验目的

1. 掌握 Hibernate 关联关系映射的基本概念，理解关联的方向和数量，重点理解双向一对多/多对一的关联关系，及其在实际应用中的体现。

2. 学习 Hibernate 框架处理一对多/多对一关联关系的方法，掌握关联关系中持久化类的实现方法，以及相应 Hibernate 映射文件的配置方案。

3. 能在实际应用中通过 Hibernate 建立正确的一对多/多对一关联关系映射，并以面向对象的方式进行数据库访问。

二、基本知识与原理

1. 客观世界中的对象往往不是孤立存在的。例如，老师与被授课的学生存在关联关系。如果已经得到某老师的实例，那么应该可以获取该老师对应的全部学生。反之，如果已经得到一个学生的实例，也应该可以访问该学生对应的老师。这种在实例之间的相互访问关系就是关联关系。Hibernate 框架可以处理各种不同的关联关系。

2. 关联的方向可分为单向关联和双向关联。

（1）单向关联：只需单向访问关联端。例如，只能通过老师访问学生，或者只能通过学生访问老师。

（2）双向关联：关联的两端可以相互访问。例如，老师和学生之间可以相互访问。

3. 除考虑关联的方向问题之外，还要考虑关联双方的数量问题，即一对一、一对多、多对一、多对多的关联关系。

4. 双向的一对多/多对一关系是现实中最为常见的关联关系。假设实体类 A 到实体类 B 是一对多（一个 A 的实例关联多个 B 的实例），则 B 到 A 就是多对一（多个 B 的实例可能关联同一个 A 的实例）。要表示这种关系，则 B 类（"多"的一端）中关联一个 A 的实例，而 A 类（"一"的一端）中关联一个集合对象，集合元素为 B 的实例。

5. 在 Hibernate 映射文件中，作为"一"的一端，需要使用<set···/>或<bag···/>元素来映射关联属性；作为"多"的一端，则需要使用<many-to-one···/>元素来映射关联属性。

三、实验内容及步骤

1. 在 MySQL 中创建一个名称为 hibernatedb 的数据库,并在该数据库中创建一个名称为 customer 的数据表,创建表的 DDL 如下所示。

```
CREATE TABLE 'customer'(
  'customerID' INTEGER(11) NOT NULL,
  'account' VARCHAR(20) DEFAULT NULL,
  'password' VARCHAR(20) DEFAULT NULL,
  'name' VARCHAR(20) DEFAULT NULL,
  'sex' TINYINT(1) DEFAULT NULL,
  'birthday' DATE DEFAULT NULL,
  'email' VARCHAR(100) DEFAULT NULL,

  PRIMARY KEY ('customerID')
)
```

2. 在表 customer 中添加 3 条记录,具体内容如表 7-1 所示。

表 7-1   customer 中的记录

用 户 编 号	登录用户名	登录密码
1	zjut	zjut
2	admin	admin
3	temp	temp

3. 在 hibernatedb 数据库中创建一个名称为 address 的数据表,用于记录用户的联系地址。customer 与 address 是一对多的关系,其中 address 表中的 cust_id 是外键,参考 customer 表的主键。创建表的 DDL 如下所示。

```
CREATE TABLE 'address'(
  'addressID' INTEGER(11) NOT NULL,
  'detail' VARCHAR(200) DEFAULT NULL,
  'zipcode' VARCHAR(10) DEFAULT NULL,
  'phone' VARCHAR(20) DEFAULT NULL,
  'type' VARCHAR(20) DEFAULT NULL,
  'cust_id' INTEGER(11) DEFAULT NULL,

  PRIMARY KEY('addressID'),
  FOREIGN KEY('cust_id') REFERENCES customer('customerID')
)
```

4. 在 Eclipse 中新建 Web 工程 hibernate-prj3,并添加 MySQL 驱动程序库文件、commons-logging-1.2.jar、Struts2 核心包和 Hibernate 核心包到工程中。

5. 在 hibernate-prj3/src 中新建配置文件 hibernate.cfg.xml,具体代码可参照实验五中基础实验里的 hibernate.cfg.xml。

6. 在 hibernate-prj3 中新建 cn.edu.zjut.po 包,并在其中创建持久化类 Customer.java 和 Address.java。Customer 与 Address 是一对多的关系,因此需要在 Customer 中增加一

个 Set 集合属性,用于记录它关联的一系列 Address 实体,而在 Address 中只需增加一个 Customer 类型的属性,代码片段如下所示。

/** Customer **/

```java
package cn.edu.zjut.po;
…
public class Customer {
    private int customerId;
    private String account;
    private String password;
    private String repassword;
    private String name;
    private Boolean sex;
    private String sexStr;
    private Date birthday;
    private String email;
    private Set addresses = new HashSet(0);
    //省略构造函数
    //省略 getters/setters 方法
}
```

/** Address **/

```java
package cn.edu.zjut.po;
…
public class Address {
    private int addressId;
    private String detail;
    private String zipcode;
    private String phone;
    private String type;
    private Customer customer;
    //省略构造函数
    //省略 getters/setters 方法
}
```

7. 在 hibernate-prj3 的 cn.edu.zjut.po 包中,新建映射文件 Customer.hbm.xml,作为"一"的一端,需要使用<set…/>或<bag…/>元素来映射关联属性。在<set…/>或<bag…/>元素中,需要增加<key…/>子元素来映射外键列,并使用<one-to-many…/>子元素映射关联属性,代码片段如下所示。

```xml
<hibernate-mapping>
    <class name = "cn.edu.zjut.po.Customer" table = "customer" catalog = "hibernatedb">
        <id name = "customerId" type = "int">
            <column name = "customerID" />
            <generator class = "increment" />
        </id>
        …
        <set name = "addresses" inverse = "true" cascade = "all" lazy = "false">
            <key column = "cust_id"/>
            <one-to-many class = "cn.edu.zjut.po.Address"/>
```

```
    </set>
  </class>
</hibernate-mapping>
```

8. 在 hibernate-prj3 的 cn.edu.zjut.po 包中,新建映射文件 Address.hbm.xml,作为"多"的一端。需要使用<many-to-one…/>元素来映射关联属性,代码片段如下所示。

```
<hibernate-mapping>
    <class name = "cn.edu.zjut.po.Address" table = "address" catalog = "hibernatedb">
        <id name = "addressId" type = "int">
            <column name = "addressID" />
            <generator class = "increment" />
        </id>
        ⋮
        <many-to-one name = "customer" class = "cn.edu.zjut.po.Customer"
                              fetch = "select" not-null = "true">
            <column name = "cust_id" />
        </many-to-one>
    </class>
</hibernate-mapping>
```

9. 修改配置文件 hibernate.cfg.xml,增加 Customer.hbm.xml 与 Address.hbm.xml 映射文件的声明。

10. 在 hibernate-prj3 中新建 cn.edu.zjut.dao 包,并在其中创建 DAO 操作辅助类 HibernateUtil.java 和数据库操作基础类 BaseHibernateDAO.java(可参考实验六中基础实验里的代码)。

11. 在 cn.edu.zjut.dao 包中创建数据库操作类 CustomerDAO.java 和 AddressDAO.java(代码略)。

12. 在 hibernate-prj3 中新建 cn.edu.zjut.service 包,在其中创建 UserService.java,并实现用户登录和增加地址的逻辑,代码片段如下所示。

```java
package cn.edu.zjut.service;
⋮
public class UserService {
    public boolean login(Customer loginUser) {
        //代码略
    }
    public boolean addAddr(Customer loginUser, Address address) {
        ActionContext ctx = ActionContext.getContext();
        request = (Map) ctx.get("request");
        CustomerDAO c_dao = new CustomerDAO();
        loginUser = (Customer)c_dao
                            .findById(loginUser.getCustomerId());
        address.setCustomer(loginUser); //注释 1
        loginUser.getAddresses().add(address);
        Transaction tran = null;
        try {
            tran = c_dao.getSession().beginTransaction();
            c_dao.update(loginUser);
```

```
            tran.commit();
            request.put("loginUser", loginUser);
            request.put("tip", "添加地址成功!");
            return true;
        } catch (Exception e) {
            if(tran != null) tran.rollback();
            return false;
        } finally {
            c_dao.getSession().close();
        }
    }
}
```

13. 在 hibernate-prj3 中新建 cn.edu.zjut.action 包,并在其中创建 UserAction.java,代码片段如下所示。

```
package cn.edu.zjut.action;
    ⋮
public class UserAction {
    private Customer loginUser;
    private Address address;
    public String login() {
        UserService userServ = new UserService();
        if (userServ.login(loginUser))
            return "success";
        else
            return "fail";
    }
    public String addAddr() {
        UserService userServ = new UserService();
        if (userServ.addAddr(loginUser, address))
            return "success";
        else
            return "fail";
    }
    //省略 getters/setters 方法
}
```

14. 在 hibernate-prj3 中新建 login.jsp 页面,作为用户登录的视图(代码略)。

15. 在 hibernate-prj3 中新建 loginSuccess.jsp 页面,作为登录成功的视图。在该视图中显示登录用户的所有个人信息(包括地址信息),并在该视图中增加"添加新地址"的表单。代码片段如下所示。

```
<table>
<tr><td>个人信息:<p></td></tr>
<tr>
    <td>用户名:</td>
    <td><s:property value="#request.loginUser.account" /></td>
</tr>
    ⋮
```

```
<s:iterator value="#request.loginUser.addresses" status="st">
<tr><td>地址<s:property value="#st.count"/>: </td><tr>
<tr><td>详细地址: </td><td><s:property value="detail" /></td><tr>
    ⋮
</s:iterator>
</table>
<hr>
添加新地址: <p>
<s:form action="UseraddAddr" method="post">
    <s:hidden name="loginUser.customerId"
              value="%{#request.loginUser.customerId}"/>
    <s:textfield name="address.detail" label="详细地址" />
    <s:textfield name="address.zipcode" label="邮政编码" />
    <s:textfield name="address.phone" label="联系电话" />
    <s:textfield name="address.type"
                 label="地址类型(office,home,etc.)" />
    <s:submit value="添加"/>
</s:form>
```

16. 在工程 hibernate-pr3 的 src 目录中创建 struts.xml 文件,用于配置 Action 并设置页面导航,使得在登录成功或添加地址成功时都转向 loginSuccess.jsp 页面(代码略)。

17. 编辑 Web 应用的 web.xml 文件,增加 Struts2 核心 Filter 的配置(代码略)。

18. 将 hibernate-prj3 部署在 Tomcat 服务器上。

19. 通过浏览器访问 login.jsp 页面,并记录运行结果。

20. 修改 hibernate.cfg.xml 配置文件,增加属性,使得能在控制台输出 SQL 语句。将 hibernate-prj3 重新部署在 Tomcat 服务器上并运行。当 Customer.hbm.xml 中 set 元素的 inverse 属性为 true 时,观察并记录控制台输出的 SQL 语句。然后,将 inverse 属性值设置成 false,观察并记录控制台输出的 SQL 语句。

21. 修改 Customer.hbm.xml 中 set 元素的 lazy 属性值为 true。将 hibernate-prj3 重新部署在 Tomcat 服务器上并运行,观察并记录 loginSuccess.jsp 页面的输出。

22. 修改 UserService.java,将"注释 1"处的代码删除。将 hibernate-prj3 重新部署在 Tomcat 服务器上并运行,观察并记录运行结果。

四、实验要求

1. 填写并上交实验报告,报告中应包括如下内容。

(1) 运行结果截图。

(2) 根据实验步骤 6~8,总结双向一对多/多对一关联关系中持久化类的实现方法以及相应 Hibernate 映射文件的配置方案,并记录下来。

(3) 查找相关资料,总结单向一对多/多对一关联关系中持久化类的实现方法以及相应 Hibernate 映射文件的配置方案,并记录下来。

(4) 根据实验步骤 20~22,查找相关资料,总结 Hibernate 映射文件的 set 元素中 inverse、lazy、cascade 等属性的取值及作用,以及 many-to-one 元素中的 not-null、fetch 等属性的作用,并记录下来。

(5) 碰到的问题及解决方案或对问题的思考。

(6) 实验收获及总结。

2. 上交程序源代码,代码中应有相关注释。

提高实验——多对多关联

一、实验目的

1. 进一步掌握 Hibernate 关联关系映射的基本概念,理解关联的方向和数量,重点理解双向多对多的关联关系及其在实际应用中的体现。

2. 学习 Hibernate 框架处理多对多关联关系的方法,掌握关联关系中持久化类的实现方法以及相应 Hibernate 映射文件的配置方案。

3. 能在实际应用中通过 Hibernate 建立正确的多对多关联关系映射,并以面向对象的方式进行数据库访问。

二、基本知识与原理

1. 多对多关系在现实中也很常见。例如,学生与选修课之间的关系。一名学生可以选择多门选修课,而每门选修课又可以被多名学生选择。

2. 数据库中的多对多关联关系一般需采用连接表的方式处理,将多对多关系转化为 2 个一对多关系。若在实体类 A 和 B 之间表示双向多对多关系,则需要在 A 类和 B 类中各关联一个集合对象,集合元素为对方的实例。

3. 在 Hibernate 映射文件中,两个"多"端都需要使用<set…/>或<bag…/>元素来映射关联属性,并在其 table 属性中指定连接表的名字。

三、实验内容及步骤

1. 在 hibernatedb 数据库中创建一个名称为 address2 的数据表,用于记录用户的联系地址。其中,customer 与 address2 是多对多的关系。创建表的 DDL 如下所示。

```
CREATE TABLE 'address2'(
  'addressID' INTEGER(11) NOT NULL,
  'detail' VARCHAR(200) DEFAULT NULL,
  'zipcode' VARCHAR(10) DEFAULT NULL,
  'phone' VARCHAR(20) DEFAULT NULL,
  'type' VARCHAR(20) DEFAULT NULL,

  PRIMARY KEY ('addressID')
)
```

2. 在 hibernatedb 数据库中创建一个名称为 join 的连接表,将 customer 与 address2 的多对多关系转化为两个一对多关系。其中,cust_id 是外键,参考 customer 表的主键;addr_id 也是外键,参考 address2 表的主键。创建表的 DDL 如下所示。

```
CREATE TABLE 'join'(
  'cust_id' INTEGER(11) NOT NULL,
  'addr_id' INTEGER(11) NOT NULL,
```

```
    PRIMARY KEY ('cust_id', addr_id),
    FOREIGN KEY ('cust_id') REFERENCES customer ('customerID')
    FOREIGN KEY ('addr_id') REFERENCES customer ('addressID')
)
```

3. 修改 cn.edu.zjut.po 包中的 Address.java。由于 Customer 与 Address 是多对多的关系,因此需要在 Address 中增加一个 Set 集合属性,用于记录它关联的一系列 Customer 实体。代码片段如下所示。

```
package cn.edu.zjut.po;
 ⋮
public class Address {
    private int addressId;
    private String detail;
    private String zipcode;
    private String phone;
    private String type;
    private Set customers = new HashSet(0);
    //省略构造函数
    //省略 getters/setters 方法
}
```

4. 修改 cn.edu.zjut.po 包中的映射文件 Customer.hbm.xml,使用＜set…/＞或＜bag…/＞元素来映射关联属性,并在其 table 属性中指定连接表的名字。代码片段如下所示。

```
<hibernate-mapping>
    <class name="cn.edu.zjut.po.Customer" table="customer" catalog="hibernatedb">
        <id name="customerId" type="int">
            <column name="customerID" />
            <generator class="increment" />
        </id>
         ⋮
        <set name="addresses" table="join" cascade="all">
            <key column="cust_id"/>
            <many-to-many column="addr_id"
                          class="cn.edu.zjut.po.Address"/>
        </set>
    </class>
</hibernate-mapping>
```

5. 修改 cn.edu.zjut.po 包中的 Address.hbm.xml,使得持久化类 Address 与数据库表 address2 相映射,并使用＜set…/＞或＜bag…/＞元素来映射关联属性,同时在其 table 属性中指定连接表的名字(代码略)。

6. 将 hibernate-prj3 重新部署在 Tomcat 服务器上。通过浏览器访问 login.jsp 页面,并记录运行结果。

7. 修改 loginSuccess.jsp 页面,在该视图中增加删除用户地址的表单。代码片段如下(字体加粗部分)所示。

```
<table>
<tr><td>个人信息:<p></td></tr>
<tr>
    <td>用户名:</td>
<td><s:property value="#request.loginUser.account"/></td>
</tr>
    ⋮
<s:iterator value="#request.loginUser.addresses" status="st">
<s:form action="UserdelAddr" method="post">
    <s:hidden name="loginUser.customerId"
              value="%{#request.loginUser.customerId}"/>
    <s:hidden name="address.addressId" value="%{addressId}"/>
    <tr><td>地址<s:property value="#st.count"/>:</td><tr>
    <tr><td>详细地址:</td><td><s:property value="detail"/></td><tr>
      ⋮
    <tr><td><s:submit value="删除"/></td><tr>
</s:form>
</s:iterator>
</table>
```

8. 修改 cn. edu. zjut. service 包中的 UserService. java,在其中添加用户删除地址的逻辑。注意,是删除用户与地址的关联,不是删除地址本身(代码略)。

9. 修改 cn. edu. zjut. action 包中的 UserAction. java,在其中添加地址删除的方法(代码略)。

10. 修改 struts. xml 文件,添加删除用户地址的页面导航(代码略)。

11. 将 hibernate-prj3 重新部署在 Tomcat 服务器上。通过浏览器访问 login. jsp 页面,并记录运行结果。

12. 修改映射文件 Customer. hbm. xml 和 Address. hbm. xml 中<set>元素的 inverse、lazy 和 cascade 属性的值,观察并记录运行结果。

四、实验要求

1. 填写并上交实验报告,报告中应包括如下内容。

(1) 运行结果截图。

(2) 根据实验步骤 3~5,总结双向多对多关联关系中持久化类的实现方法以及相应 Hibernate 映射文件的配置方案,并记录下来。

(3) 查找相关资料,总结单向多对多关联关系中持久化类的实现方法以及相应 Hibernate 映射文件的配置方案,并记录下来。

(4) 根据实验过程,查找相关资料,总结 Hibernate 映射文件的<set>元素中 inverse、lazy、cascade 等属性的取值及作用,以及这些属性在多对多关联关系和一对多/多对一关联关系中取值方法的异同,并记录下来。

(5) 碰到的问题及解决方案或对问题的思考。

(6) 实验收获及总结。

2. 上交程序源代码,代码中应有相关注释。

扩展实验——一对一关联

一、实验目的

1. 进一步掌握 Hibernate 关联关系映射的基本概念,理解关联的方向和数量,重点理解双向一对一的关联关系及其在实际应用中的体现。

2. 学习 Hibernate 框架处理一对一关联关系的方法,掌握关联关系中持久化类的实现方法以及相应 Hibernate 映射文件的配置方案。

3. 能在实际应用中通过 Hibernate 建立正确的一对一关联关系映射,并以面向对象的方式进行数据库访问。

二、基本知识与原理

1. 在现实中的一对一关系,例如学生与身份证之间的关系。一名学生只有一份身份证信息,而一个身份证也只能对应一名学生。

2. Hibernate 中的一对一关联分为主键关联和外键关联两类。主键关联是通过主表和辅表的主键相关联,即两个表的主键值是一样的,不必增加额外的字段;而在外键关联方式里,主表和辅表通过外键相关联,可以将其看成是一对多/多对一关联关系的特殊形式,其中的多方退化成了一方。

3. 若采用外键关联,则在 Hibernate 映射文件中,"一"方将使用＜one-to-one＞元素来映射关联属性,并在其中用 property-ref 指定关联类的属性名。而"多"方若要退化成"一"方,只需要在＜many-to-one＞元素中设置"unique"="true"即可。若采用主键关联,则在 Hibernate 映射文件中,两方都使用＜one-to-one＞元素来映射关联属性,并将辅表的＜id＞元素(主键)的＜generator＞(生成策略)取值改成 foreign,同时加上属性参数。

三、实验内容及步骤

1. 修改 hibernatedb 数据库里的 customer 表与 address 表,使它们形成一对一外键关联的关系,即一个用户只能与一个地址形成唯一的对应关系。

2. 修改 cn.edu.zjut.po 包中的 Customer.java 和 Address.java。由于本例中 Customer 与 Address 是一对一的关系,因此 Customer 中包括一个 Address 类型的属性,Address 中包括一个 Customer 类型的属性。代码片段如下所示。

```
/** Customer **/
package cn.edu.zjut.po;
...
public class Customer {
    private int customerId;
    private String account;
    private String password;
    private String repassword;
    private String name;
    private Boolean sex;
```

```
    private String sexStr;
    private Date birthday;
    private String email;
    private Address address;
    //省略构造函数
    //省略 getters/setters 方法
}
```

/** Address **/
```
package cn.edu.zjut.po;
    ⋮
public class Address {
    private int addressId;
    private String detail;
    private String zipcode;
    private String phone;
    private String type;
    private Customer customer;
    //省略构造函数
    //省略 getters/setters 方法
}
```

3. 修改 cn.edu.zjut.po 包中的映射文件 Customer.hbm.xml 和 Address.hbm.xml，使它们形成一对一外键关联的关系，即，在 Customer.hbm.xml 中使用＜one-to-one…/＞元素映射关联属性，并在其中用 property-ref="customer" 来指明对应关联类中用于保存本类的属性名；而作为一对多/多对一关联关系的特例，在 Address.hbm.xml 中仍使用＜many-to-one…/＞元素来映射关联属性，然后用属性 unique="true" 做限制。代码片段如下所示。

/** Customer.hbm.xml **/
```
<hibernate-mapping>
    <class name="cn.edu.zjut.po.Customer" table="customer" catalog="hibernatedb">
        <id name="customerId" type="int">
            <column name="customerID" />
            <generator class="increment" />
        </id>
        ⋮
        <one-to-one name="address" class="cn.edu.zjut.po.Address"
                    cascade="all" property-ref="customer"/>
    </class>
</hibernate-mapping>
```

/** Address.hbm.xml **/
```
<hibernate-mapping>
    <class name="cn.edu.zjut.po.Address" table="address" catalog="hibernatedb">
        <id name="addressId" type="int">
            <column name="addressID" />
            <generator class="increment" />
```

```
        </id>
        ⋮
        <many-to-one name="customer" class="cn.edu.zjut.po.Customer"
                     fetch="select" unique="true">
            <column name="cust_id"/>
        </many-to-one>
    </class>
</hibernate-mapping>
```

4. 修改 loginSuccess.jsp 页面,使得在该视图中,若用户已有一个地址信息,则将其显示出来,并能选择删除该地址信息;若用户没有地址信息,则显示添加新地址的表单。代码片段如下所示。

```
<table>
<tr><td>个人信息:<p></td></tr>
<tr>
    <td>用户名:</td>
    <td><s:property value="#request.loginUser.account"/></td>
</tr>
 ⋮
<s:if test="#request.loginUser.address">
    <s:form action="UserdelAddr" method="post">
    <s:hidden name="loginUser.customerId"
              value="%{#request.loginUser.customerId}"/>
    <tr>
          <td>详细地址:</td>
       <td><s:property
                  value="#request.loginUser.address.detail"/></td>
    <tr>
     ⋮
    <tr><td><s:submit value="删除"/></td><tr>
    </s:form>
</s:if>
<s:else>
    添加新地址:<p>
    <s:form action="UseraddAddr" method="post">
    <s:hidden name="loginUser.customerId"
              value="%{#request.loginUser.customerId}"/>
    <s:textfield name="address.detail" label="详细地址"/>
     ⋮
    <s:submit value="添加"/>
    </s:form>
</s:else>
</table>
```

5. 对 cn.edu.zjut.service 包中的 UserService.java 和 cn.edu.zjut.action 包中的 UserAction.java 做相应的修改,使得用户能够添加或删除一个地址(代码略)。

6. 将 hibernate-prj3 重新部署在 Tomcat 服务器上。通过浏览器访问 login.jsp 页面,并记录运行结果。

7. 尝试修改映射文件中<one-to-one⋯/>或<many-to-one⋯/>元素的 cascade 等属

性的值,观察并记录运行结果。

8. 在 hibernatedb 数据库中创建一个名称为 address3 的数据表,用于记录用户的联系地址。其中 customer 与 address3 是一对一主键关联的关系,创建表的 DDL 如下所示。

```
CREATE TABLE 'address3'(
  'addressID' INTEGER(11) NOT NULL,
  'detail' VARCHAR(200) DEFAULT NULL,
  'zipcode' VARCHAR(10) DEFAULT NULL,
  'phone' VARCHAR(20) DEFAULT NULL,
  'type' VARCHAR(20) DEFAULT NULL,

  PRIMARY KEY ('addressID'),
)
```

9. 维持 cn.edu.zjut.po 包中的 Customer.java 和 Address.java 代码不变,修改映射文件 Customer.hbm.xml 和 Address.hbm.xml,使它们形成一对一主键关联的关系,即,双方都使用<one-to-one…/>元素映射关联属性,在辅表 Addres.hbm.xml 的<one-to-one…/>元素中添加属性 constrained="true"表示受到外键约束,并将其主键的<generator>设置成 foreign,同时增加属性参数。代码片段如下所示。

```
/** Customer.hbm.xml **/
<hibernate-mapping>
    <class name="cn.edu.zjut.po.Customer" table="customer" catalog="hibernatedb">
        <id name="customerId" type="int">
            <column name="customerID"/>
            <generator class="increment"/>
        </id>
        ⋮
        <one-to-one name="address" class="cn.edu.zjut.po.Address"/>
    </class>
</hibernate-mapping>


/** Address.hbm.xml **/
<hibernate-mapping>
    <class name="cn.edu.zjut.po.Address" table="address" catalog="hibernatedb">
        <id name="addressId" type="int">
            <column name="addressID"/>
            <generator class="foreign">
                <param name="property">customer</param>
            </generator>
        </id>
        ⋮
        <one-to-one name="customer" class="cn.edu.zjut.po.Customer"
                    cascade="all" constrained="true"/>
    </class>
</hibernate-mapping>
```

10. 将 hibernate-prj3 重新部署在 Tomcat 服务器上。通过浏览器访问 login.jsp 页面,

并记录运行结果。

11. 尝试修改映射文件中＜one-to-one…/＞元素的 cascade 等属性的值,观察并记录运行结果。

四、实验要求

1. 填写并上交实验报告,报告中应包括如下内容。

(1) 运行结果截图。

(2) 根据实验过程,总结双向一对一关联关系中持久化类的实现方法,以及主键关联和外键关联情况下 Hibernate 映射文件的配置方案,并记录下来。

(3) 查找相关资料,总结单向一对一关联关系中持久化类的实现方法以及相应 Hibernate 映射文件的配置方案,并记录下来。

(4) 根据实验过程,查找相关资料,总结一对一关联关系下,Hibernate 映射文件的＜one-to-one…/＞元素 cascade 等属性的取值及作用,并记录下来。

(5) 碰到的问题及解决方案或对问题的思考。

(6) 实验收获及总结。

2. 上交程序源代码,代码中应有相关注释。

实验八

SSH 整合 (Spring4+Struts2+Hibernate4) ——基于 SSH 的用户注册模块

基础实验——Spring 框架搭建

一、实验目的

1. 掌握 Spring 环境搭建的基本方法,能在 JaveSE 应用中使用 Spring,并能在 Eclipse 中开发 Spring 应用。

2. 初步理解 Spring 的核心机制:控制反转 IoC(Inversion of Control)与依赖注入 DI (Dependency Injection)。

3. 理解 Spring 配置文件的作用,掌握 bean 元素及其属性的作用和基本配置方法。

二、基本知识与原理

1. Spring 为企业应用的开发提供了一个轻量级的解决方案。

2. Spring 框架共包括 7 个模块,每个模块分别用于提供不同的解决方案,具体如下。

(1) Spring Core 模块:提供控制反转(IoC)容器,是 Spring 框架的核心机制,其他特性都基于 IoC 之上。

(2) Spring Context 模块:提供对 Spring 中对象的框架式访问方式。

(3) Spring DAO 模块:提供了集成 JDBC 的封装包。

(4) Spring ORM 模块:提供了集成常用 ORM 框架的封装包。

(5) Spring Web 模块:提供了 Web 开发以及集成 Web 框架的封装包。

(6) Spring AOP 模块:提供了面向切面编程(AOP)的实现。

(7) Spring MVC 模块:提供了一个 MVC 框架。

3. 在传统的程序设计过程中,当某个 Java 实例(调用者)需要调用另一个 Java 实例(被调用者)时,通常由调用者来创建被调用者的实例;而在控制反转模式下,创建被调用者的工作不再由调用者来完成,两者之间的依赖关系将由 Spring 管理,以使两者解耦。

4. 在 Spring 中,创建被调用者的工作由 Spring 容器来完成。然后再将被调用者实例注入调用者,因此该过程也被称为依赖注入。

5. Spring 推荐面向接口编程,这样可以更好地让规范和实现分离,从而提供更好的

实验八　SSH 整合(Spring4+Struts2+Hibernate4)——基于 SSH 的用户注册模块

解耦。

三、实验内容及步骤

1. 登录 http://maven.springframework.org/release/org/springframework/spring/ 站点,下载 Spring 框架的依赖 JAR 包(如:spring-framework-4.0.0.RELEASE-dist)。

2. 在 Eclipse 中新建 Java 工程 spring-prj1,并添加 common-logging-1.2.jar 和 Spring 的 4 个基础 JAR 包到工程中,如图 8-1 所示。

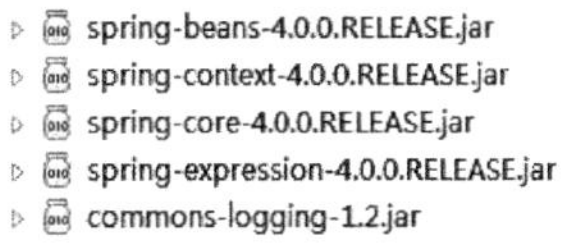

- spring-beans-4.0.0.RELEASE.jar
- spring-context-4.0.0.RELEASE.jar
- spring-core-4.0.0.RELEASE.jar
- spring-expression-4.0.0.RELEASE.jar
- commons-logging-1.2.jar

图 8-1　Spring 的 4 个基础包

3. 在 spring-prj1 中新建 cn.edu.zjut.dao 包,并在其中创建 ICustomerDAO 接口定义数据持久层的操作以及实现类 CustomerDAO 实现数据持久层的操作,具体代码如下所示。

```java
package cn.edu.zjut.dao;

public interface ICustomerDAO {
    public void save();
}
```

```java
package cn.edu.zjut.dao;

public class CustomerDAO implements ICustomerDAO{
    public CustomerDAO(){
        System.out.println("create CustomerDao.");
    }
    public void save() {
        System.out.println("execute -- save() -- method.");
    }
}
```

4. 在 spring-prj1 中创建 Spring 配置文件 applicationContext.xml,并在其中配置 CustomerDAO 实例,具体代码如下所示。

```xml
<?xml version = "1.0" encoding = "UTF-8"?>
<beans xmlns = "http://www.springframework.org/schema/beans"
    xmlns:xsi = "http://www.w3.org/2001/XMLSchema-instance"
    xmlns:p = "http://www.springframework.org/schema/p"
    xsi:schemaLocation = "http://www.springframework.org/schema/beans http://www.springframework.org/schema/beans/spring-beans-4.0.xsd">

    <bean id = "userDAO" class = "cn.edu.zjut.dao.CustomerDAO" />

</beans>
```

5. 在 spring-prj1 中新建 cn.edu.zjut.app 包,并在其中创建测试类 SpringEnvTest,调用 CustomerDAO 实例的 save()方法,具体代码如下所示。

```java
package cn.edu.zjut.app;
```

```java
import org.springframework.context.ApplicationContext;
import org.springframework.context.support.ClassPathXmlApplicationContext;
import cn.edu.zjut.dao.ICustomerDAO;

public class SpringEnvTest {
    public static void main(String[] args) {
        //创建 Spring 容器
        ApplicationContext ctx = new ClassPathXmlApplicationContext(
                "applicationContext.xml");
        //获取 CustomerDAO 实例
        ICustomerDAO userDao = (ICustomerDAO) ctx.getBean("userDAO");
        userDao.save();
    }
}
```

6. 运行测试类 SpringEnvTest，观察控制台的输出，并记录运行结果。

7. 在 spring-prj1 中新建 cn.edu.zjut.service 包，并在其中创建 IUserService 接口（定义注册逻辑）以及实现类 UserService（实现注册逻辑），具体代码如下所示。

```java
package cn.edu.zjut.service;

public interface IUserService {
    public void register();
}
```

```java
package cn.edu.zjut.service;
import cn.edu.zjut.dao.ICustomerDAO;

public class UserService implements IUserService {
    private ICustomerDAO customerDAO = null;

    public UserService(){
        System.out.println("create UserService.");
    }
    public void setCustomerDAO(ICustomerDAO customerDAO) {
        System.out.println(" -- setCustomerDAO -- ");
        this.customerDAO = customerDAO;
    }
    public void register() {
        System.out.println("execute -- register() -- method.");
        customerDAO.save();
    }
}
```

8. 修改 Spring 配置文件 applicationContext.xml，在其中增加对 userService 实例的配置，具体代码如下所示。

```xml
<?xml version = "1.0" encoding = "UTF-8"?>
<beans xmlns = "http://www.springframework.org/schema/beans"
    xmlns:xsi = "http://www.w3.org/2001/XMLSchema-instance"
    xmlns:p = "http://www.springframework.org/schema/p"
```

实验八　SSH整合(Spring4+Struts2+Hibernate4)——基于SSH的用户注册模块

```
        xsi:schemaLocation = "http://www.springframework.org/schema/beans http://www.
springframework.org/schema/beans/spring-beans-4.0.xsd">

    <bean id = "userDAO" class = "cn.edu.zjut.dao.CustomerDAO" />

    <bean id = "userService" class = "cn.edu.zjut.service.UserService">
        <property name = "customerDAO" ref = "userDAO" />
    </bean>

</beans>
```

9. 修改测试类 SpringEnvTest,调用 userService 实例的 register()方法,具体代码如下所示。

```
package cn.edu.zjut.app;
import org.springframework.context.ApplicationContext;
import org.springframework.context.support.ClassPathXmlApplicationContext;
import cn.edu.zjut.dao.ICustomerDAO;

public class SpringEnvTest {
    public static void main(String[] args) {
        //创建 Spring 容器
        ApplicationContext ctx = new ClassPathXmlApplicationContext(
                "applicationContext.xml");
        //获取 UserService 实例
        IUserService userService =
                    (IUserService) ctx.getBean("userService");
        userService.register();
    }
}
```

10. 运行测试类 SpringEnvTest,观察控制台的输出,并记录运行结果。

四、实验要求

1. 填写并上交实验报告,报告中应包括如下内容。
(1) 运行结果截图。
(2) 根据实验过程,观察运行后的控制台输出,查找相关资料,总结 Spring 容器管理 Bean 组件的过程(如：何时加载、何时调用 Bean 实例中的方法等)以及进行依赖注入的过程,并记录下来。
(3) 根据实验步骤 4、8,查找相关资料,总结 Spring 配置文件中 bean 元素及其属性、子元素的作用,并记录下来。
(4) 根据实验步骤 5、9,查找相关资料,总结控制反转模式下两个 Java 实例的依赖关系与传统的程序设计过程体现出来的依赖关系有什么区别,控制反转模式的优点是什么,并将其记录下来。
(5) 碰到的问题及解决方案或对问题的思考。
(6) 实验收获及总结。
2. 上交程序源代码,代码中应有相关注释。

提高实验——Spring 与 Hibernate 的整合

一、实验目的

1. 进一步熟悉 Spring 基础环境搭建的方法,以及在 Eclipse 中开发 Spring 应用的主要步骤。

2. 掌握 Spring 框架与 Hibernate 框架整合的基本步骤,理解 Spring 容器对 DataSource 实例和 SessionFactory 实例的管理方式。

3. 进一步理解 Spring 中控制反转 IoC 的核心机制。

4. 进一步熟悉 Spring 配置文件,掌握 Spring 配置文件中对 DataSource 和 SessionFactory 的配置方法和属性注入的方式。

二、基本知识与原理

1. Spring 框架与 Hibernate 框架整合。将生成 DataSource 对象和 SessionFactory 对象的过程交给 Spring 容器实现,而不在代码中实现,即由 IoC 容器控制对象的生成和属性的注入。

2. 要使用 IoC 装配对象(如配置 SessionFactory 对象),就必须在 Spring 配置文件(默认为 applicationContext.xml)中进行配置。

3. Spring 框架中的 IoC 容器管理的对象都被称为 bean。bean 都需要在配置文件的 <beans> 元素下使用 <bean> 元素配置。

4. Spring IoC 容器的代表者是 API 中的 BeanFactory 接口。所有 IoC 容器装配成功的对象,都将通过 BeanFactory 获得,进而在应用中得到使用。

三、实验内容及步骤

1. 在 MySQL 中创建一个名称为 hibernatedb 的数据库,并在该数据库中创建一个名称为 customer 的数据表,表结构如表 8-1 所示。

表 8-1 customer 数据表

字 段 名 称	类　　型	中 文 含 义
customerID	INTEGER(11),Primary key,Not Null	用户编号
account	VARCHAR(20)	登录用户名
password	VARCHAR(20)	登录密码
name	VARCHAR(20)	真实姓名
sex	BOOLEAN(1)	性别
birthday	DATE	出生日期
phone	VARCHAR(20)	联系电话
email	VARCHAR(100)	电子邮箱
address	VARCHAR(200)	联系地址
zipcode	VARCHAR(10)	邮政编码
fax	VARCHAR(20)	传真号码

2. 在 Java 工程 spring-prj1 中添加 MySQL 驱动程序库文件和 Hibernate 核心包,并添加 Spring 框架中与数据库操作相关的 3 个 JAR 包:spring-jdbc-4.0.0.RELEASE.jar、spring-orm-4.0.0.RELEASE.jar、spring-tx-4.0.0.RELEASE.jar。

3. 在 spring-prj1 中新建 cn.edu.zjut.po 包,并在其中创建持久化类 Customer.java 以及 Hibernate 映射文件 Customer.hbm.xml,具体代码可参照实验五中基础实验里的内容。

4. 在 cn.edu.zjut.dao 包中创建数据库操作基础类 BaseHibernateDAO.java,具体代码如下所示。

```java
package cn.edu.zjut.dao;
import org.hibernate.Session;
import org.hibernate.SessionFactory;

public class BaseHibernateDAO{
    private SessionFactory sessionFactory;

    public Session getSession(){
        return sessionFactory.openSession();
    }

    public void setSessionFactory(SessionFactory sessionFactory) {
        this.sessionFactory = sessionFactory;
    }
}
```

5. 修改 cn.edu.zjut.dao 包中的 ICustomerDAO 接口和 CustomerDAO 实现类,为 save()方法增加一个 Customer 类型的输入参数,并使 CustomerDAO 继承数据库操作基础类 BaseHibernateDAO,具体代码如下所示。

```java
package cn.edu.zjut.dao;
import cn.edu.zjut.po.Customer;

public interface ICustomerDAO {
    void save(Customer transientInstance);
}
package cn.edu.zjut.dao;
import org.hibernate.Session;
import org.hibernate.Transaction;
import cn.edu.zjut.po.Customer;

public class CustomerDAO extends BaseHibernateDAO
                         implements ICustomerDAO{

    public void save(Customer transientInstance) {
        Transaction tran = null;
        Session session = null;
        try {
            session = getSession();
            tran = session.beginTransaction();
            session.save(transientInstance);
```

```
            tran.commit();
        } catch (RuntimeException re) {
            if(tran != null) tran.rollback();
            throw re;
        } finally {
            session.close();
        }
    }
}
```

6. 相应地,修改 cn.edu.zjut.service 包中的 IUserService 接口和 UserService 实现类,为 register()方法增加一个 Customer 类型的输入参数,代码略。

7. 修改 Spring 配置文件 applicationContext.xml,修改 beans 元素的属性,并在其中增加对数据源的配置,代码片段如下所示。

```xml
<?xml version = "1.0" encoding = "UTF-8"?>
<beans xmlns = "http://www.springframework.org/schema/beans"
    xmlns:xsi = "http://www.w3.org/2001/XMLSchema-instance"
    xmlns:context = "http://www.springframework.org/schema/context"
    xmlns:aop = "http://www.springframework.org/schema/aop"
    xmlns:tx = "http://www.springframework.org/schema/tx"
    xsi:schemaLocation = "http://www.springframework.org/schema/beans

http://www.springframework.org/schema/beans/spring-beans-4.0.xsd
        http://www.springframework.org/schema/aop

http://www.springframework.org/schema/aop/spring-aop-4.0.xsd
        http://www.springframework.org/schema/tx

http://www.springframework.org/schema/tx/spring-tx-4.0.xsd
        http://www.springframework.org/schema/context

http://www.springframework.org/schema/context/spring-context-4.0.xsd">

    <bean id = "dataSource" class = "org.springframework.jdbc.
    datasource.DriverManagerDataSource">
        <property name = "driverClassName"
                value = "com.mysql.jdbc.Driver"/>
        <property name = "url"
value = "jdbc:mysql://localhost:3306/hibernatedb"/>
        <property name = "username" value = "root"/>
        <property name = "password" value = ""/>
    </bean>
    ⋮
</beans>
```

8. 修改 Spring 配置文件 applicationContext.xml,在其中增加对 SessionFactory 实例的配置,代码片段如下所示。

```xml
<bean id = "sessionFactory"
class = "org.springframework.orm.hibernate4.LocalSessionFactoryBean">
```

```xml
<property name="dataSource" ref="dataSource"/>
<property name="hibernateProperties">
    <props>
        <prop key="hibernate.dialect">
            org.hibernate.dialect.MySQLDialect
        </prop>
    </props>
</property>
<property name="mappingResources">
    <list>
        <value>cn/edu/zjut/po/Customer.hbm.xml</value>
    </list>
</property>
</bean>
```

9. 修改 Spring 配置文件 applicationContext.xml,在其中增加对 BaseHibernateDAO 实例的配置,并在其中注入 sessionFactory,同时修改 CustomerDAO 实例的配置属性,代码片段如下所示。

```xml
<bean id="baseDAO" class="cn.edu.zjut.dao.BaseHibernateDAO">
    <property name="sessionFactory" ref="sessionFactory"/>
</bean>

<bean id="userDAO" class="cn.edu.zjut.dao.CustomerDAO"
    parent="baseDAO"/>
```

10. 修改测试类 SpringEnvTest,注册一个新用户,具体代码如下所示。

```java
package cn.edu.zjut.app;
import org.springframework.context.ApplicationContext;
import org.springframework.context.support.ClassPathXmlApplicationContext;
import cn.edu.zjut.po.Customer;
import cn.edu.zjut.service.IUserService;

public class SpringEnvTest {
    public static void main(String[] args) {
        ApplicationContext ctx = new ClassPathXmlApplicationContext(
                "applicationContext.xml");
        IUserService userService =
                (IUserService) ctx.getBean("userService");
        Customer cust = new Customer();
        cust.setAccount("SPRING");
        cust.setPassword("SPRING");
        userService.register(cust);
    }
}
```

11. 运行测试类 SpringEnvTest,记录运行结果。

四、实验要求

1. 填写并上交实验报告,报告中应包括如下内容。
(1) 运行结果截图。

（2）根据实验过程，总结 DataSource、SessionFactory、CustomerDAO、UserService 对象之间的依赖关系，并记录下来。

（3）根据实验步骤 7~9，查找相关资料，总结 Spring 配置文件中 DataSource、SessionFactory、CustomerDAO、UserService 的配置方法以及属性注入的方式。

（4）碰到的问题及解决方案或对问题的思考。

（5）实验收获及总结。

2. 上交程序源代码，代码中应有相关注释。

扩展实验——Spring、Struts 与 Hibernate 的整合

一、实验目的

1. 掌握 Spring 框架与 Hibernate 框架、Struts 框架整合的基本步骤，理解 Spring 容器对 Bean 实例的管理方式。

2. 理解 Spring 容器对 Struts2 核心控制器 Action 的管理方式，并理解 Struts 配置文件和 web.xml 中产生的相应变化。

3. 进一步理解 Spring 中控制反转 IoC 的核心机制。

二、基本知识与原理

1. 使用 Spring 整合 Struts2 框架，其核心思想是将 Struts2 的 Action 实例交给 Spring 框架的 IoC 容器装配、管理。因此，在 Action 类中，应提供必要的 setters 方法以注入所需的属性。同时，struts.xml 文件中的＜action＞元素的 class 属性将不再是该 Action 对应的实际类型，而是与 applicationContext.xml 中 Action 的 bean 的 id 对应。

2. 使用 Spring 整合 Struts2 框架，还需要在 web.xml 文件中配置一个 listener 来完成加载 Spring 配置文件的功能。

三、实验内容及步骤

1. 在 Eclipse 中新建 Web 工程 spring-prj1，并将 common-logging-1.2.jar、MySQL 驱动程序库文、Hibernate 核心包和 Spring 的 7 个 JAR 包（其中包括 Spring 的 4 个基础 JAR 包以及与数据库操作相关的 3 个 JAR 包）添加到工程中。

2. 将 Struts2 中的 8 个核心包（参考实验二中的基础实验步骤）以及 Struts2 对 Spring 进行支持的 JAR 包 struts2-spring-plugin-2.3.15.1.jar 添加到工程中。

3. 将 Spring 支持 web 开发的 JAR 包 spring-web-4.0.0.RELEASE.jar 添加到工程中。

4. 在 spring-prj1 中新建 cn.edu.zjut.po 包，并在其中创建持久化类 Customer.java 以及 Hibernate 映射文件 Customer.hbm.xml，代码略。

5. 在 spring-prj1 中新建 cn.edu.zjut.dao 包，并在其中创建数据库操作基础类 BaseHibernateDAO.java、ICustomerDAO 接口和 CustomerDAO 实现类，代码略。

6. 在 spring-prj1 中新建 cn.edu.zjut.service 包，并在其中创建 IUserService 接口和

UserService 实现类，代码略。

7. 在 spring-prj1 中新建 cn.edu.zjut.action 包，并在其中创建 UserAction.java，用于调用用户注册逻辑，代码片段如下所示。

```java
package cn.edu.zjut.action;
import cn.edu.zjut.po.Customer;
import cn.edu.zjut.service.IUserService;

public class UserAction {
    private Customer loginUser;
    private IUserService userService = null;

    //省略 loginUser 的 getters/setters 方法

    public void setUserService(IUserService userService) {
        this.userService = userService;
    }

    public String execute() {
        userService.register(loginUser);
        return "success";
    }
}
```

8. 在项目的 WebRoot/WEB-INF/路径下创建 Spring 配置文件 applicationContext.xml，并参考实验八提高实验中的内容进行配置。

9. 修改 Spring 配置文件 applicationContext.xml，增加对 UserAction 实例的配置，代码片段如下所示。

```xml
<bean id="userAction" class="cn.edu.zjut.action.UserAction"
    scope="prototype">
    <property name="userService" ref="userService" />
</bean>
```

10. 在项目的 src/路径下创建 Struts2 配置文件 struts.xml，用于配置 Action 并设置页面导航，请注意其中<action>元素的 class 属性取值，具体代码如下所示。

```xml
<?xml version="1.0" encoding="UTF-8" ?>
<!DOCTYPE struts PUBLIC "-//Apache Software Foundation//DTD Struts Configuration 2.1//EN"
"http://struts.apache.org/dtds/struts-2.1.dtd">
<struts>
    <package name="strutsBean" extends="struts-default"
            namespace="/">
        <action name="register" class="userAction">
            <result name="success">/regSuccess.jsp</result>
            <result name="fail">/regFail.jsp</result>
        </action>
    </package>
</struts>
```

11. 编辑 Web 应用的 web.xml 文件,增加对 Struts2 核心 Filter 的配置,并添加对 Spring 监听器的配置,代码片段如下所示。

```xml
<web-app>
  <listener>
      <listener-class>
              org.springframework.web.context.ContextLoaderListener
      </listener-class>
  </listener>
  ⋮
<web-app>
```

12. 在 spring-prj1 中新建 register.jsp 页面,作为用户注册视图;新建 regiSuccess.jsp 和 regFail.jsp 页面,分别作为注册成功和失败的视图(代码略)。

13. 将 spring-prj1 部署在 Tomcat 服务器上。

14. 通过浏览器访问 register.jsp 页面,并记录运行结果。

四、实验要求

1. 填写并上交实验报告,报告中应包括如下内容。
(1) 运行结果截图。
(2) 结合实验过程,总结 Spring 整合 Struts2 框架的关键步骤,并记录下来。
(3) 根据实验步骤 7,总结本实验中的 UserAction 与以往实验中的写法的关键区别,并记录下来。
(4) 根据实验步骤 9,查找相关资料,总结配置文件 applicationContext.xml 中 bean 元素的 prototype 属性及其取值的含义,并记录下来。
(5) 根据实验步骤 11,查找相关资料,总结 web.xml 文件中添加监听器的目的,并记录下来。
(6) 碰到的问题及解决方案或对问题的思考。
(7) 实验收获及总结。

2. 上交程序源代码,代码中应有相关注释。

实验九

Spring的核心机制：控制反转(IoC)
——登录用户的购物车

基础实验——Spring 容器中的依赖注入

一、实验目的

1. 进一步掌握 Spring 环境搭建的基本方法，能熟练地在 Java SE 应用中使用 Spring，并能熟练地在 Eclipse 中开发 Spring 应用。

2. 深入理解 Spring 的核心机制：控制反转 IoC(Inversion of Control)与依赖注入 DI(Dependency Injection)。

3. 深入理解 Spring 配置文件的作用，掌握配置文件中各主要元素及其属性的作用和基本配置方法。

二、基本知识与原理

1. Spring 中主要有 2 种注入方法：设置注入和构造器注入。

2. 设置注入是指 IoC 容器使用属性的 setter 方法来注入被依赖的实例。这种注入方式简单、直观，因而在 Spring 的依赖注入里得到了频繁使用。

3. 使用设置注入的配置文件，在 bean 元素下用 property 元素指定属性名。其中，property 元素的 name 属性值必须与对应的 setter 方法名相对应，这样才能调用 setter 方法注入具体值。

4. 构造器注入是指 IoC 容器通过调用带参的构造方法注入所依赖的属性。这种方式在构造实例时已经为其完成了依赖关系的初始化。

5. 使用构造器注入的配置文件，在 bean 元素下用 constructor-arg 元素表示构造方法的参数。其中，constructor-arg 元素的 index 属性表示构造方法中参数的索引值。

6. 无论是设置注入还是构造器注入，都要为方法指定具体的参数值，为属性赋值。参数值有不同的类型，可以分为 3 种情况：基本数据类型和 String 类型、其他 bean 类型、null 值。

7. 当类的属性是集合类型时，也可以使用 IoC 进行注入。常用的集合类型有 List、Set、Map 以及 Properties，相应的配置文件可以使用<list>、<set>、<map>和<props>元素进行配置。

三、实验内容及步骤

1. 在 Eclipse 中新建 Java 工程 spring-prj2,并添加 common-logging-1.2.jar 和 Spring 的 4 个基础 JAR 包到工程中(可参考实验八基础实验中的实验步骤 2)。

2. 在 spring-prj2 中新建 cn.edu.zjut.bean 包,并在其中创建 IItem 接口(代码略)及其实现类 Item。代码片段如下所示。

```
package cn.edu.zjut.bean;

public class Item implements IItem{

    private String itemID;
    private String title;
    private String description;
    private double cost;

    public Item(String itemID, String title,
            String description, double cost) {
        this.itemID = itemID;
        this.title = title;
        this.description = description;
        this.cost = cost;
        System.out.println("create Item.");
    }
     //省略 getters/setters 方法
}
```

3. 在 spring-prj2 中创建 Spring 配置文件 applicationContext.xml,并在其中使用构造器注入的方式配置 Item 实例,具体代码如下所示。

```xml
<?xml version = "1.0" encoding = "UTF-8"?>
< beans xmlns = "http://www.springframework.org/schema/beans"
    xmlns:xsi = "http://www.w3.org/2001/XMLSchema-instance"
    xmlns:p = "http://www.springframework.org/schema/p"
    xsi:schemaLocation = "http://www.springframework.org/schema/beans http://www.springframework.org/schema/beans/spring-beans-4.0.xsd">

    < bean id = "item1" class = "cn.edu.zjut.bean.Item">
        < constructor-arg index = "0" type = "java.lang.String">
            < value >978-7-121-12345-1</value >
        </constructor-arg >
        < constructor-arg index = "1" type = "java.lang.String">
            < value >Java EE 技术实验指导教程</value >
        </constructor-arg >
        < constructor-arg index = "2" type = "java.lang.String">
            < value >Web 程序设计知识回顾、轻量级 Java EE 应用框架、企业级 EJB 组件编程技术、Java EE 综合应用开发</value >
        </constructor-arg >
        < constructor-arg index = "3" type = "double">
```

```xml
            <value>19.95</value>
        </constructor-arg>
    </bean>

    <bean id="item2" class="cn.edu.zjut.bean.Item">
        <constructor-arg index="0" type="java.lang.String">
            <value>978-7-121-12345-2</value>
        </constructor-arg>
        <constructor-arg index="1" type="java.lang.String">
            <value>Java EE 技术</value>
        </constructor-arg>
        <constructor-arg index="2" type="java.lang.String">
            <value>Struts 框架、Hibernate 框架、Spring 框架、会话 Bean、实体 Bean、消息驱动 Bean</value>
        </constructor-arg>
        <constructor-arg index="3" type="double">
            <value>29.95</value>
        </constructor-arg>
    </bean>

</beans>
```

4. 在 spring-prj2 中新建 cn.edu.zjut.app 包,并在其中创建测试类 SpringEnvTest,对构造器注入进行测试,具体代码如下所示。

```java
package cn.edu.zjut.app;
import org.springframework.context.ApplicationContext;
import org.springframework.context.support.ClassPathXmlApplicationContext;
import cn.edu.zjut.dao.ICustomerDAO;

public class SpringEnvTest {
    public static void main(String[] args) {
            ApplicationContext ctx = new ClassPathXmlApplicationContext(
                "applicationContext.xml");
        IItem item1 = (IItem) ctx.getBean("item1");
        System.out.println(item1.getItemID());
        System.out.println(item1.getTitle());
        System.out.println(item1.getDescription());
        System.out.println(item1.getCost());
        IItem item2 = (IItem) ctx.getBean("item2");
        System.out.println(item2.getItemID());
        System.out.println(item2.getTitle());
        System.out.println(item2.getDescription());
        System.out.println(item2.getCost());
    }
}
```

5. 运行测试类 SpringEnvTest,观察控制台的输出,并记录运行结果。

6. 在 spring-prj2 的 cn.edu.zjut.bean 包中创建 IItemOrder 接口(代码略)及其实现类 ItemOrder,代码片段如下所示。

```java
package cn.edu.zjut.bean;

public class ItemOrder implements IItemOrder {
    private IItem item;
    private int numItems;

    public void incrementNumItems() {
        setNumItems(getNumItems() + 1);
    }
    public void cancelOrder() {
        setNumItems(0);
    }
    public double getTotalCost() {
        return (getNumItems() * getUnitCost());
    }
    //省略 getters/setters 方法
}
```

7. 修改 Spring 配置文件 applicationContext.xml,在其中使用设置注入的方式增加 ItemOrder 实例的配置,代码片段如下所示。

```xml
<bean id="itemorder1" class="cn.edu.zjut.bean.ItemOrder">
    <property name="numItems"><value>1</value></property>
    <property name="item"><ref bean="item1"/></property>
</bean>

<bean id="itemorder2" class="cn.edu.zjut.bean.ItemOrder">
    <property name="numItems"><value>2</value></property>
    <property name="item"><ref bean="item2"/></property>
</bean>
```

8. 修改测试类 SpringEnvTest,对构造器注入进行测试,代码片段如下所示。

```java
public class SpringEnvTest {
    public static void main(String[] args) {
        ApplicationContext ctx = new ClassPathXmlApplicationContext(
            "applicationContext.xml");
        IItemOrder itemorder1 = (IItemOrder) ctx.getBean("itemorder1");
        System.out.println("书名:" + itemorder1.getItem().getTitle());
        System.out.println("数量:" + itemorder1.getNumItems());
        IItemOrder itemorder2 = (IItemOrder) ctx.getBean("itemorder2");
        System.out.println("书名:" + itemorder2.getItem().getTitle());
        System.out.println("数量:" + itemorder2.getNumItems());
    }
}
```

9. 运行测试类 SpringEnvTest,观察控制台的输出,并记录运行结果。

10. 在 spring-prj2 的 cn.edu.zjut.bean 包中创建购物车类 ShoppingCart.java,其中包含集合类型(List)属性,具体代码如下所示。

```java
package cn.edu.zjut.bean;
```

```
import java.util.*;

public class ShoppingCart {
    private List itemsOrdered;

    public List getItemsOrdered() {
        return (itemsOrdered);
    }
    public void setItemsOrdered(List itemsOrdered) {
        this.itemsOrdered = itemsOrdered;
    }
}
```

11. 修改 Spring 配置文件 applicationContext.xml，在其中增加对 ShoppingCart 实例的配置，并使用 list 元素对 List 类型属性进行配置，代码片段如下所示。

```xml
<bean id="shoppingcart" class="cn.edu.zjut.bean.ShoppingCart">
    <property name="itemsOrdered">
        <list>
            <ref bean="itemorder1"/>
            <ref bean="itemorder2"/>
        </list>
    </property>
</bean>
```

12. 修改测试类 SpringEnvTest，对集合类型属性配置进行测试，代码略。
13. 运行测试类 SpringEnvTest，观察控制台的输出，并记录运行结果。
14. 对其他 3 种集合类型(Set、Map、Properties)进行配置和测试，并记录运行结果。

四、实验要求

1. 填写并上交实验报告，报告中应包括如下内容。
(1) 运行结果截图。
(2) 根据实验过程，比较并总结设置注入与构造器注入各自的优点及适用的场景，并记录下来。
(3) 根据实验过程，查找相关资料，总结在设置注入或构造器注入方式下，配置文件中相应的配置方法，以及所使用的相关元素及其属性的作用，并记录下来。
(4) 根据实验步骤 10~14，查找相关资料，总结集合类型属性的配置方法，并记录下来。
(5) 碰到的问题及解决方案或对问题的思考。
(6) 实验收获及总结。
2. 上交程序源代码，代码中应有相关注释。

提高实验——Spring 容器中的 Bean

一、实验目的

1. 理解 Spring IoC 容器中 bean 的 5 个作用域所起到的作用，以及使用不同作用域的

bean 之间的区别。

2. 掌握 Spring IoC 容器创建 bean 实例的 3 种方式。

3. 掌握 Spring 在 bean 的依赖关系注入之后执行特定行为的方法,会通过使用 init-method 属性的方式以及让 Bean 类实现 InitializingBean 接口的方式,添加 bean 的依赖关系注入之后的行为。

4. 掌握 Spring 在 bean 销毁之前执行特定行为的方法,会通过使用 destroy-method 属性的方式以及让 Bean 类实现 DisposableBean 接口的方式,添加 bean 销毁之前的行为。

5. 理解 Spring IoC 容器对 bean 生命周期的管理方式。

二、基本知识与原理

1. Spring 框架的 IoC 容器不仅可以向 bean 注入不同的依赖属性,还可以指定其作用域,用以告诉容器如何生成该 bean。bean 有 5 种作用域,分别是 singleton、prototype、request、session 和 global session。

2. Spring IoC 容器创建 bean 实例的方法主要有以下 3 种。

(1) 调用构造器创建 bean 实例。

(2) 调用静态工厂方法创建 bean 实例。

(3) 调用实例工厂方法创建 bean 实例。

3. Spring 可以管理 singleton 作用域 bean 的生命周期,并允许在 bean 的依赖关系注入之后以及 bean 销毁之前执行特定的行为。

4. 管理 bean 的依赖关系注入之后的行为可以有以下 2 种方式。

(1) 使用 init-method 属性指定,某个方法应在 bean 全部依赖关系设置结束后自动执行。

(2) 让 Bean 类实现 InitializingBean 接口,并使用该接口提供的 afterPropertiesSet 方法。

5. 管理 bean 销毁之前的行为也可以有以下 2 种方式。

(1) 使用 destroy-method 属性指定,某个方法应在 bean 销毁之前自动执行。

(2) 让 Bean 类实现 DisposableBean 接口,并使用该接口提供的 destroy 方法。

三、实验内容及步骤

1. 修改 Spring 配置文件 applicationContext.xml,为 Item 实例的配置添加作用域属性(scope),代码片段如下所示。

```xml
<bean id = "item1" class = "cn.edu.zjut.bean.Item" scope = "singleton">
    <constructor-arg index = "0" type = "java.lang.String">
        <value>978-7-121-12345-1</value>
    </constructor-arg>
    ⋮
</bean>

<bean id = "item2" class = "cn.edu.zjut.bean.Item" scope = "prototype">
    <constructor-arg index = "0" type = "java.lang.String">
        <value>978-7-121-12345-2</value>
```

```
        </constructor-arg>
    ...
</bean>
```

2. 修改测试类 SpringEnvTest,对 bean 的作用域进行测试,代码片段如下所示。

```java
public class SpringEnvTest {
    public static void main(String[] args) {
            ApplicationContext ctx = new ClassPathXmlApplicationContext(
                "applicationContext.xml");
        System.out.println("getBean(item1) --- 1");
        IItem item11 = (IItem) ctx.getBean("item1");
        System.out.println("getBean(item1) --- 2");
        IItem item12 = (IItem) ctx.getBean("item1");
        System.out.println("getBean(item2) --- 1");
        IItem item21 = (IItem) ctx.getBean("item2");
        System.out.println("getBean(item2) --- 2");
        IItem item22 = (IItem) ctx.getBean("item2");
    }
}
```

3. 运行测试类 SpringEnvTest,观察控制台的输出,并记录运行结果。

4. 修改 cn.edu.zjut.bean 包中的 ItemOrder.java,在其构造方法及 setters 方法中添加相应的输出,代码片段如下所示。

```java
package cn.edu.zjut.bean;

public class ItemOrder implements IItemOrder {
    private IItem item;
    private int numItems;

    public ItemOrder() {
        System.out.println("Spring 实例化 ItemOrder …");
    }
    public void setItem(IItem item) {
        System.out.println("Spring 注入 item …");
        this.item = item;
    }
    public void setNumItems(int n) {
        System.out.println("Spring 注入 numItems …");
        this.numItems = n;
    }
    ...
}
```

5. 修改测试类 SpringEnvTest,对 bean 的初始化方法进行测试,代码片段如下所示。

```java
public class SpringEnvTest {
    public static void main(String[] args) {
            ApplicationContext ctx = new ClassPathXmlApplicationContext(
                "applicationContext.xml");
        IItemOrder itemorder1 = (IItemOrder)ctx.getBean("itemorder1");
```

6. 运行测试类 SpringEnvTest，观察控制台的输出，并记录运行结果。

7. 修改 cn.edu.zjut.bean 包中的 ItemOrder.java，使其实现 InitializingBean 接口，并通过实现 InitializingBean 接口的方式，添加 bean 的依赖关系注入之后的行为，代码片段如下所示。

```
package cn.edu.zjut.bean;

public class ItemOrder implements IItemOrder, InitializingBean {
    public void afterPropertiesSet() throws Exception {
        System.out.println("正在执行初始化方法 afterPropertiesSet…");
    }
    ⋮
}
```

8. 运行测试类 SpringEnvTest，观察控制台的输出，并记录运行结果。

9. 修改 cn.edu.zjut.bean 包中的 ItemOrder.java，通过自定义方法，添加 bean 的依赖关系注入之后的行为，代码片段如下所示。

```
package cn.edu.zjut.bean;

public class ItemOrder implements IItemOrder, InitializingBean {
    public void init() {
        System.out.println("正在执行初始化方法 init…");
    }
    ⋮
}
```

10. 修改 Spring 配置文件 applicationContext.xml，通过使用 init-method 属性的方式，为 ItemOrder 实例添加 bean 的依赖关系注入之后的行为，代码片段如下所示。

```xml
<bean id="itemorder1" class="cn.edu.zjut.bean.ItemOrder"
      init-method="init">
    <property name="numItems"><value>1</value></property>
    <property name="item"><ref bean="item1"/></property>
</bean>
```

11. 运行测试类 SpringEnvTest，观察控制台的输出，并记录运行结果。

12. 修改 cn.edu.zjut.bean 包中的 ItemOrder.java，通过实现 DisposableBean 接口的方式，添加 bean 销毁之前的行为（代码略）。

13. 运行测试类 SpringEnvTest，观察控制台的输出，并记录运行结果。

14. 修改 cn.edu.zjut.bean 包中的 ItemOrder.java，通过自定义方法，添加 bean 销毁之前的行为（代码略）；并修改 Spring 配置文件 applicationContext.xml，通过使用 destroy-method 属性的方式，为 ItemOrder 实例添加 bean 销毁之前的行为（代码略）。

15. 运行测试类 SpringEnvTest，观察控制台的输出，并记录运行结果。

16. 在工程 spring-prj2 的 cn.edu.zjut.bean 包中创建工厂类 ItemOrderFactory.java，

并在类中提供静态工厂方法,返回 ItemOrder 实例,具体代码如下所示。

```
package cn.edu.zjut.bean;

public class ItemOrderFactory {
    public static ItemOrder createItemOrder() {
        System.out.println("调用静态工厂方法创建 bean…");
        return new ItemOrder();
    }
}
```

17. 修改 Spring 配置文件 applicationContext.xml,使用 factory-method 属性调用静态工厂方法,创建 ItemOrder 实例,代码片段如下所示。

```xml
<bean id="itemorder2" class="cn.edu.zjut.bean.ItemOrderFactory"
    factory-method="createItemOrder">
    <property name="numItems"><value>1</value></property>
    <property name="item"><ref bean="item1"/></property>
</bean>
```

18. 修改测试类 SpringEnvTest,对静态工厂方法创建 bean 的方法进行测试,代码片段如下所示。

```java
public class SpringEnvTest {
    public static void main(String[] args) {
        ApplicationContext ctx = new ClassPathXmlApplicationContext(
            "applicationContext.xml");
        IItemOrder itemorder2 = (IItemOrder)ctx.getBean("itemorder2");
    }
}
```

19. 运行测试类 SpringEnvTest,观察控制台的输出,并记录运行结果。

20. 修改工厂类 ItemOrderFactory.java,将其中的静态工厂方法修改为实例工厂方法,具体代码如下所示。

```java
package cn.edu.zjut.bean;

public class ItemOrderFactory {
    public ItemOrder createItemOrder() {
        System.out.println("调用实例工厂方法创建 bean…");
        return new ItemOrder();
    }
}
```

21. 修改 Spring 配置文件 applicationContext.xml,使用 factory-method 属性和 factory-bean 属性调用实例工厂方法,创建 ItemOrder 实例,代码片段如下所示。

```xml
<bean id="itemorderFactory"
    class="cn.edu.zjut.bean.ItemOrderFactory"/>
<bean id="itemorder2" class="cn.edu.zjut.bean.ItemOrder"
    factory-method="createItemOrder"
```

```
                factory-bean="itemorderFactory">
    <property name="numItems"><value>1</value></property>
    <property name="item"><ref bean="item1"/></property>
</bean>
```

22. 运行测试类 SpringEnvTest，观察控制台的输出，并记录运行结果。

四、实验要求

1. 填写并上交实验报告，报告中应包括如下内容。
（1）运行结果截图。
（2）根据实验步骤 1～3，查找相关资料，总结 Spring IoC 容器中 bean 的 5 个作用域（singleton、prototype、request、session、global session）以及使用不同作用域的 bean 之间的区别，并记录下来。
（3）根据实验步骤 4～6，查找相关资料，总结 Spring IoC 容器管理的 bean 的生命周期，并记录下来。
（4）根据实验步骤 7～15，查找相关资料，总结 Spring 如何管理在 bean 的依赖关系注入之后以及 bean 销毁之前执行特定的行为，总结 Spring 执行 Bean 实例化、依赖关系注入、初始化方法和析构方法的顺序，并将其记录下来。
（5）根据实验步骤 4～6、16～22，查找相关资料，总结 Spring IoC 容器创建 bean 实例的 3 种方式及相应的配置过程，并记录下来。
（6）碰到的问题及解决方案或对问题的思考。
（7）实验收获及总结。

2. 上交程序源代码，代码中应有相关注释。

扩展实验——深入 Spring 容器

一、实验目的

1. 理解 Spring 中 BeanFactory 的作用，掌握使用 XmlBeanFactory 创建 BeanFactory 实例的方法，掌握使用 FileSystemXmlApplicationContext 或 ClassPathXmlApplicationContext 创建 ApplicationContext 实例的方法。

2. 理解 Spring 中 ApplicationContext 对国际化功能的支持，并掌握进行国际化的基本步骤。

3. 理解 Spring 的事件机制，掌握 Spring 完成事件监听及处理的基本步骤。

二、基本知识与原理

1. Spring 容器最基本的接口是 BeanFactory。BeanFactory 负责配置、创建、管理 bean，包括管理 bean 与 bean 之间的依赖关系。BeanFactory 有一个实现类 org.springframework.beans.factory.xml.XmlBeanFactory。

2. ApplicationContext 是 BeanFactory 的子接口，对于大部分 Java EE 应用而言，使用它作为 Spring 容器更方便。ApplicationContext 的常用实现类是 FileSystemXmlApplicationContext

实验九 Spring 的核心机制：控制反转(IoC)——登录用户的购物车

和 ClassPathXmlApplicationContext。

3. ApplicationContext 接口继承 MessageSource 接口，用以实现国际化功能。

4. Spring 支持事件机制，通过 ApplicationEvent 类和 ApplicationListener 接口，可以实现 ApplicationContext 的事件处理。

三、实验内容及步骤

1. 修改测试类 SpringEnvTest，分别使用 XmlBeanFactory 创建 BeanFactory 实例，或使用 FileSystemXmlApplicationContext、ClassPathXmlApplicationContext 创建 ApplicationContext 实例，代码片段如下所示。

```java
public class SpringEnvTest {
    public static void main(String[] args) {
        FileSystemResource isr = new FileSystemResource("src/applicationContext.xml");
        XmlBeanFactory factory = new XmlBeanFactory(isr);
        IItemOrder itemorder3 = (IItemOrder) factory.getBean("itemorder3");
    }
}
public class SpringEnvTest {
    public static void main(String[] args) {
        ClassPathResource res = new ClassPathResource("applicationContext.xml");
        XmlBeanFactory factory = new XmlBeanFactory(res);
        ⋮
    }
}
public class SpringEnvTest {
    public static void main(String[] args) {
        ApplicationContext ctx = new ClassPathXmlApplicationContext(
            "applicationContext.xml");
        IItemOrder itemorder3 = (IItemOrder)ctx.getBean("itemorder3");
    }
}
public class SpringEnvTest {
    public static void main(String[] args) {
        ApplicationContext ctx = new FileSystemXmlApplicationContext(
            "src/applicationContext.xml");
        ⋮
    }
}
```

2. 运行测试类 SpringEnvTest，观察控制台的输出，并记录运行结果。

3. 在 spring-prj2 中新建 cn.edu.zjut.local 包，并在其中创建简体中文的资源文件 message_zh_CN.properties，在其中输入"HelloWorld={0}，现在是{1}"，并对其进行转码，最终得到如下结果。

```
HelloWorld = {0}\uFF0C\u73B0\u5728\u662F{1}
```

4. 在 cn.edu.zjut.local 包中创建美式英语的资源文件 message_en_US.properties，具体代码如下所示。

```
HelloWorld = {0},now is{1}
```

5. 修改 Spring 配置文件 applicationContext.xml，增加名为 messageSource 的 bean 实例，以完成国际化的配置，代码片段如下所示。

```
<bean id = "messageSource" class =
"org.springframework.context.support.ResourceBundleMessageSource">
    <property name = "basenames">
        <list>
            <value>local/message</value>
            <!-- 如果有多个资源文件,全部列在此处 -->
        </list>
    </property>
</bean>
```

6. 修改测试类 SpringEnvTest，对 ApplicationContext 对国际化功能的支持进行测试，代码片段如下所示。

```
public class SpringEnvTest {
    public static void main(String[] args) {
            ApplicationContext ctx = new ClassPathXmlApplicationContext(
                "applicationContext.xml");
            Object[] objects = new Object[]{"HelloWorld",new Date()};
            String message = ctx.getMessage("HelloWorld",objects,Locale.CHINA);
            //String message = ctx.getMessage("HelloWorld",objects,Locale.US);
            System.out.println(message);
    }
}
```

7. 运行测试类 SpringEnvTest，观察控制台的输出，并记录运行结果。

8. 修改测试类 SpringEnvTest，将其中的"Locale.CHINA"替换成"Locale.US"。运行测试类 SpringEnvTest，观察控制台的输出，并记录运行结果。

9. 添加 spring-aop-4.0.0.RELEASE.jar 到工程 spring-prj2 中。

10. 在 spring-prj2 中新建 cn.edu.zjut.event 包，在其中创建 Spring 容器事件类 EmailEvent，并继承 ApplicationEvent 类，具体代码如下所示。

```
package cn.edu.zjut.event;
import org.springframework.context.ApplicationEvent;

public class EmailEvent extends ApplicationEvent{
    private String address;
    private String text;

    public EmailEvent(Object source){
        super(source);
    }
    public EmailEvent(Object source, String address, String text){
        super(source);
        this.setAddress(address);
```

```
        this.setText(text);
    }
    //省略 getters/setters 方法
}
```

11. 在 spring-prj2 中新建 cn.edu.zjut.listener 包,并在其中创建 Spring 容器事件的监听器类,并实现 ApplicationListener 接口,具体代码如下所示。

```
package cn.edu.zjut.listener;
import org.springframework.context.ApplicationEvent;
import org.springframework.context.ApplicationListener;
import cn.edu.zjut.event.EmailEvent;

public class EmailNotifier implements ApplicationListener{
    public void onApplicationEvent(ApplicationEvent evt){
        if(evt instanceof EmailEvent){
            EmailEvent emailEvent = (EmailEvent)evt;
            System.out.println("需要发送邮件的接收地址 " +
                                    emailEvent.getAddress());
            System.out.println("需要发送邮件的邮件正文 " +
                                    emailEvent.getText());
        }
        else{
            System.out.println("容器本身的事件 " + evt);
        }
    }
}
```

12. 修改 Spring 配置文件 applicationContext.xml,使用 Spring 容器事件注册监听器,代码片段如下所示。

```
<bean class = "cn.edu.zjut.listener.EmailNotifier"/>
```

13. 修改测试类 SpringEnvTest,调用 ApplicationContext 的 publishEvent 来触发事件,代码片段如下所示。

```
public class SpringEnvTest {
    public static void main(String[] args) {
        ApplicationContext ctx = new ClassPathXmlApplicationContext(
                "applicationContext.xml");
        EmailEvent ele = new EmailEvent("hello",
                "spring_test@zjut.edu.cn", "this is a test");
        ctx.publishEvent(ele);
    }
}
```

14. 运行测试类 SpringEnvTest,观察控制台的输出,并记录运行结果。

四、实验要求

1. 填写并上交实验报告,报告中应包括如下内容。
(1) 运行结果截图。

（2）结合实验步骤 1，总结使用 XmlBeanFactory 创建 BeanFactory 实例，或使用 FileSystemXmlApplicationContext、ClassPathXmlApplicationContext 创建 ApplicationContext 实例的基本方法，并记录下来。

（3）根据实验步骤 3～8，总结 ApplicationContext 支持国际化功能的基本步骤，并记录下来。

（4）根据实验步骤 9～14，查找相关资料，总结 Spring 完成事件监听及处理的基本步骤，并记录下来。

（5）碰到的问题及解决方案或对问题的思考。

（6）实验收获及总结。

2. 上交程序源代码，代码中应有相关注释。

Spring的面向切面编程(AOP)
——用户登录模块的增强处理

基础实验——使用@AspectJ实现AOP

一、实验目的

1. 学习理解AOP(Aspect Oriented Programming)即面向切面编程的基本概念,重点掌握切面、增强处理、切点的概念。

2. 掌握使用@AspectJ实现Spring AOP的基本步骤和配置方法,会使用基于Annotation的注解方式或基于XML配置文件的方式来定义切入点和增强处理。

二、基本知识与原理

1. AOP(Aspect Oriented Programming),即面向切面编程,作为面向对象编程的一种补充,用于处理系统中分布于各个模块中共同关注的服务问题,如事务管理、安全检查、缓存、对象池管理等。

2. 面向切面编程的相关术语主要有如下几个。

(1) 切面(Aspect):业务流程运行的某个特定步骤,也就是应用运行过程的关注点。关注点可能横切多个对象,因此也被称为横切关注点。

(2) 连接点(Joinpoint):程序执行过程中明确的点,如方法的调用或异常的抛出。Spring AOP中,连接点总是方法的调用。

(3) 增强处理(Advice):是切面的具体实现,即在前面的某个特定的连接点上执行动作。Spring中执行的动作往往就是调用某类的具体方法,如实现保存日志功能的类就是通知。

(4) 切入点(Pointcut):可以插入增强处理的连接点。如果某连接点将被添加增强处理,则该连接点就变成了切入点。

3. 在@AspectJ方式下,Spring有以下2种途径来定义切入点和增强处理。

(1) 基于Annotation的注解方式:使用@Aspect、@Pointcut等Annotation来标注切入点和增强处理。

(2) 基于XML配置文件的方式:使用Sping配置文件来定义切入点和增强处理。

三、实验内容及步骤

1. 在 Eclipse 中新建 Java 工程 spring-prj3，并添加 common-logging-1.2.jar 和 Spring 的 4 个基础 JAR 包到工程中（可参考实验八基础实验中的实验步骤 2）。

2. 在工程 spring-prj3 中添加 Spring 框架中与 AOP 相关的 JAR 包 spring-aop-4.0.0.RELEASE.jar。

3. 在站点 http://www.eclipse.org/aspectj/downloads.php 下载 aspectj-1.7.1.jar，并添加其中的 aspectjrt.jar、aspectjweaver.jar 到工程 spring-prj3 中。

4. 在站点 http://sourceforge.net/projects/aopalliance/ 下载 aopalliance.jar，并将其添加到工程 spring-prj3 中。

5. 在 spring-prj3 中新建 cn.edu.zjut.po 包，并在其中创建持久化类 Customer.java，代码片段如下所示。

```java
package cn.edu.zjut.po;

public class Customer implements java.io.Serializable {
    private int customerId;
    private String account;
    private String password;
    ⋮
//省略 getters/setters 方法
}
```

6. 在 spring-prj3 中新建 cn.edu.zjut.dao 包，并在其中创建 ICustomerDAO 接口，定义数据持久层的操作（代码略），再创建实现类 CustomerDAO，实现数据持久层的操作，具体代码如下所示。

```java
package cn.edu.zjut.dao;
import cn.edu.zjut.po.Customer;

public class CustomerDAO implements ICustomerDAO{
    public void save(Customer transientInstance) {
        System.out.println("execute -- save() -- method.");
    }
    public void delete(Customer transientInstance) {
        System.out.println("execute -- delete() -- method.");
    }
    public void update(Customer transientInstance) {
        System.out.println("execute -- update() -- method.");
    }
    public void findByHQL(String hql) {
        System.out.println("execute -- findByHQL() -- method.");
    }
}
```

7. 在 spring-prj3 中新建 cn.edu.zjut.service 包，并在其中创建 IUserService 接口（代码略）以及实现类 UserService，代码片段如下所示。

```java
package cn.edu.zjut.service;
import cn.edu.zjut.dao.ICustomerDAO;
import cn.edu.zjut.po.Customer;

public class UserService implements IUserService {
    private ICustomerDAO customerDAO = null;

    public void setCustomerDAO(ICustomerDAO customerDAO) {
        System.out.println("call setCustomerDAO in UserService.");
        this.customerDAO = customerDAO;
    }

    public void addUser(Customer cust) {
        System.out.println("execute -- addUser() -- method.");
        customerDAO.save(cust);
    }
    public void delUser(Customer cust) { … }
    public void updateUser(Customer cust) { … }
    public void findUser(Customer cust) { … }
}
```

8. 在 spring-prj3 中创建 Spring 配置文件 applicationContext.xml，并在其中配置 CustomerDAO 实例和 UserService 实例，代码略。

9. 在 spring-prj3 中新建 cn.edu.zjut.app 包，并在其中创建测试类 SpringEnvTest，调用 UserService 实例的 addUser() 方法，具体代码如下所示。

```java
package cn.edu.zjut.app;
⋮
public class SpringEnvTest {
    public static void main(String[] args) {
        ApplicationContext ctx = new ClassPathXmlApplicationContext(
                "applicationContext.xml");
        IUserService userService = (IUserService)
                ctx.getBean("userService");
        Customer cust = new Customer();
        cust.setAccount("SPRING");
        cust.setPassword("SPRING");
        userService.addUser(cust);
    }
}
```

10. 运行测试类 SpringEnvTest，观察控制台的输出，并记录运行结果。

11. 在 spring-prj3 中新建 cn.edu.zjut.aspect 包，并在其中创建 SecurityHandler.java，用于实现权限检查，并用基于 Annotation 的注解方式定义切面。其中，用 @Aspect 修饰切面类，用 @Pointcut 定义切点，用 @Before 定义 Before 增强处理，具体代码如下所示。

```java
package cn.edu.zjut.aspect;
import org.aspectj.lang.annotation.Before;
import org.aspectj.lang.annotation.Aspect;
import org.aspectj.lang.annotation.Pointcut;
```

```java
@Aspect
public class SecurityHandler {
    /** 定义 Pointcut,Pointcut 的名称是 modify,
     * 此方法不能有返回值和参数,该方法只是一个标识 */
    @Pointcut("execution( * add * (..)) || execution( * del * (..)) || execution( * update * (..))")
    private void modify(){};

    /** 定义 Advice,标识在那个切入点何处织入此方法 */
    @Before("modify()")
    private void checkSecurity() {
        System.out.println(" --- checkSecurity() --- "); }
}
```

12. 修改 Spring 配置文件 applicationContext.xml,在头文件中添加"xmlns:aop"的命名申明,并在"xsi:schemaLocation"中指定 aop 配置的 schema 的地址,同时增加对 SecurityHandler 实例的配置,并启动注解配置 AOP 支持,具体代码如下所示。

```xml
<?xml version = "1.0" encoding = "UTF - 8"?>
<beans xmlns = "http://www.springframework.org/schema/beans"
    xmlns:xsi = "http://www.w3.org/2001/XMLSchema - instance" xmlns:p = "http://www.springframework.org/schema/p"
    xmlns:aop = "http://www.springframework.org/schema/aop"
    xsi:schemaLocation = "http://www.springframework.org/schema/beans
    http://www.springframework.org/schema/beans/spring - beans - 4.0.xsd
    http://www.springframework.org/schema/aop
    http://www.springframework.org/schema/aop/spring - aop - 4.0.xsd">

    <!-- 启动使用注解配置 AOP 支持 -->
    <aop:aspectj - autoproxy />

    <bean id = "securityHandler"
        class = "cn.edu.zjut.aspect.SecurityHandler" />
    ⋮
</beans>
```

13. 运行测试类 SpringEnvTest,观察控制台的输出,并记录运行结果。

14. 在 spring-prj3 的 cn.edu.zjut.aspect 包中创建 SecurityHandler2.java,用于实现权限检查,具体代码如下所示。

```java
package cn.edu.zjut.aspect;

public class SecurityHandler2 {
    private void checkSecurity() {
        System.out.println(" --- checkSecurity()2 --- "); }
}
```

15. 修改 Spring 配置文件 applicationContext.xml,增加对 SecurityHandler2 实例的配置,并使用 XML 配置文件的方式定义切面,代码片段如下所示。

```xml
<bean id = "securityHandler2"
      class = "cn.edu.zjut.aspect.SecurityHandler2" />
<!-- 配置文件的方式 -->
<aop:config>
    <aop:aspect id = "security" ref = "securityHandler2">
        <aop:pointcut id = "modify"
                expression = "execution(* *.add*(..))" />
        <aop:before method = "checkSecurity" pointcut-ref = "modify" />
    </aop:aspect>
</aop:config>
```

16. 运行测试类 SpringEnvTest,观察控制台的输出,并记录运行结果。

四、实验要求

1. 填写并上交实验报告,报告中应包括如下内容。
(1) 运行结果截图。
(2) 根据实验过程,查找相关资料,整理 Spring AOP 中的基本概念(如切面、增强处理、切点等),并记录下来。
(3) 根据实验过程,总结 Spring AOP 的基本步骤以及使用基于 Annotation 的注解方式或基于 XML 配置文件的方式来定义切入点和增强处理的基本方法,并记录下来。
(4) 根据实验步骤 11 或 15,查找 AspectJ 切入点表达式的相关资料,记录其中 @Pointcut 注解中切入点表达式的含义。若切入点是 cn.edu.zjut.service 包中所有实现类的增、删、改方法,思考切入点表达式的写法,并将其记录下来。
(5) 对比实验步骤 10 与 13,观察运行后的控制台输出以及相应的程序代码,总结 Spring AOP 的优点或作用、适用场景,并记录下来。
(6) 碰到的问题及解决方案或对问题的思考。
(7) 实验收获及总结。
2. 上交程序源代码,代码中应有相关注释。

提高实验——使用 Spring AOP 实现事务管理

一、实验目的

1. 进一步理解 AOP 的基本概念与作用。
2. 理解 Spring 框架基于 AOP 实现声明式事务管理的基本机制,掌握进行声明式事务管理的基本方法。
3. 掌握 Spring 配置文件为实现声明式事务管理所涉及的主要元素及属性,理解其作用,并能进行正确的配置。
4. 了解 Spring 事务管理中的 7 种事务传播行为(propagation),并理解不同的事务传播行为起到的不同作用。

二、基本知识与原理

1. 事务管理是企业应用中非常重要的部分。Spring 框架对事务管理进行了高层次的抽象,定义了各种类型的事务管理器,用来实现事务管理功能。

2. Spring 框架支持编程式事务管理,也就是可以通过编写代码的方法实现事务管理。Spring 同时也支持声明式事务管理。声明式事务管理基于 Spring AOP 实现,它不在源文件中编写代码管理事务,而是使用 AOP 框架,在 IoC 容器中进行装配。

3. Spring 配置文件中提供了＜tx：advice…/＞元素来配置事务增强处理,并使用＜aop：advisor…/＞为 IoC 容器中的 Bean 配置自动事务代理。

三、实验内容及步骤

1. 在 MySQL 中创建一个名称为 hibernatedb 的数据库,并在该数据库中创建一个名称为 customer 的数据表(表结构参照实验八中提高实验的步骤1)。

2. 在工程 spring-prj3 中添加 MySQL 驱动程序库文件和 Hibernate 核心包,并添加 Spring 框架中与数据库操作相关的 3 个 JAR 包:spring-jdbc-4.0.0.RELEASE.jar、spring-orm-4.0.0.RELEASE.jar、spring-tx-4.0.0.RELEASE.jar。

3. 在 spring-prj3 的 cn.edu.zjut.po 包中创建 Hibernate 映射文件 Customer.hbm.xml(代码略)。

4. 在工程的 cn.edu.zjut.dao 包中创建数据库操作基础类 BaseHibernateDAO.java,具体代码如下所示。

```java
package cn.edu.zjut.dao;
import org.hibernate.Session;
import org.hibernate.SessionFactory;

public class BaseHibernateDAO{
    private SessionFactory sessionFactory;

    public Session getSession(){
        return sessionFactory.getCurrentSession();
    }

    public void setSessionFactory(SessionFactory sessionFactory) {
        this.sessionFactory = sessionFactory;
    }
}
```

5. 修改 cn.edu.zjut.dao 包中的 CustomerDAO 实现类,使之继承数据库操作基础类 BaseHibernateDAO,并实现数据库操作,代码片段如下所示。

```java
package cn.edu.zjut.dao;
import org.hibernate.Query;
import org.hibernate.Session;
import cn.edu.zjut.po.Customer;

public class CustomerDAO extends BaseHibernateDAO
                implements ICustomerDAO{
    public void save(Customer transientInstance) {
        System.out.println("execute -- save() -- method.");
```

```
            try {
                getSession().save(transientInstance);
            } catch (RuntimeException re) {
                throw re;
            }
        }
        ⋮
}
```

6. 修改 Spring 配置文件 applicationContext.xml，在其中增加数据源的配置、增加 SessionFactory 实例的配置、增加 BaseHibernateDAO 实例的配置，并在其中注入 sessionFactory，同时修改 CustomerDAO 实例的配置属性，代码略。

7. 运行测试类 SpringEnvTest，观察数据库中是否新增加了 SPRING 用户，思考导致该结果的原因，并将其记录下来。

8. 修改 Spring 配置文件 applicationContext.xml，在头文件中添加"xmlns:tx"的命名申明，在"xsi:schemaLocation"中指定 tx 配置的 schema 的地址，代码片段如下所示。

```xml
<?xml version = "1.0" encoding = "UTF-8"?>
<beans xmlns = "http://www.springframework.org/schema/beans"

xmlns:context = "http://www.springframework.org/schema/context"
    xmlns:xsi = "http://www.w3.org/2001/XMLSchema-instance"
    xmlns:tx = "http://www.springframework.org/schema/tx"
    xmlns:aop = "http://www.springframework.org/schema/aop"
    xsi:schemaLocation = "
        http://www.springframework.org/schema/beans

http://www.springframework.org/schema/beans/spring-beans-4.0.xsd
        http://www.springframework.org/schema/context
http://www.springframework.org/schema/context/spring-context-4.0.xsd
        http://www.springframework.org/schema/aop

http://www.springframework.org/schema/aop/spring-aop-4.0.xsd
        http://www.springframework.org/schema/tx

http://www.springframework.org/schema/tx/spring-tx-4.0.xsd
        ">
    ⋮
</beans>
```

9. 修改 Spring 配置文件，配置事务管理器，代码片段如下所示。

```xml
<!-- 配置事务管理器 -->
<bean id = "transactionManager"
class = "org.springframework.orm.hibernate4.HibernateTransactionManager">
    <property name = "sessionFactory">
        <ref bean = "sessionFactory" />
    </property>
</bean>
```

10. 修改 Spring 配置文件,通过 AOP 的方式为工程 spring-prj3 添加事务管理,代码片段如下所示。

```xml
<!-- 定义增强处理拦截方法 -->
<tx:advice id="txAdvice"
transaction-manager="transactionManager">
    <tx:attributes>
        <tx:method name="add*" propagation="REQUIRED" />
        <tx:method name="update*" propagation="REQUIRED" />
        <tx:method name="delete*" propagation="REQUIRED" />
        <tx:method name="del*" propagation="REQUIRED" />
        <tx:method name="*" read-only="true" />
    </tx:attributes>
</tx:advice>
```

11. 修改 Spring 配置文件,定义事务拦截切面,通过 AOP 的方式添加事务管理,代码片段如下所示(为避免 Spring 中混用@Aspectj 和 Schema-based AOP 引起错误,注意删除基础实验中的 AOP 配置)。

```xml
<!-- 定义事务拦截切面 -->
<aop:config>
    <aop:pointcut id="allServiceMethod"
          expression="execution(* cn.edu.zjut.service.*.*(..))" />
    <aop:advisor pointcut-ref="allServiceMethod"
                 advice-ref="txAdvice" />
</aop:config>
```

12. 运行测试类 SpringEnvTest,观察数据库中是否新增加了 SPRING 用户,思考导致该结果的原因,并将其记录下来。

四、实验要求

1. 填写并上交实验报告,报告中应包括如下内容。

(1) 运行结果截图。

(2) 根据实验步骤 4,查找相关资料,总结 sessionFactory 对象的 openSession()方法与 getCurrentSession()方法的区别,思考在本次实验中为什么使用后者,并记录下来。

(3) 根据实验步骤 7,观察运行结果(数据库中是否新增加了 SPRING 用户),思考导致该结果的原因,并记录下来。

(4) 根据实验步骤 9~11,查找相关资料,总结 Spring 配置文件中与实现声明式事务管理相关的主要元素及其属性的作用,并记录下来。

(5) 根据实验步骤 10,查找相关资料,总结 Spring 事务管理中的 7 种事务传播行为(propagation)所起到的不同作用,并记录下来。

(6) 碰到的问题及解决方案或对问题的思考。

(7) 实验收获及总结。

2. 上交程序源代码,代码中应有相关注释。

扩展实验——Spring AOP 的核心工作原理：代理和代理工厂

一、实验目的

1. 进一步理解 AOP 的基本概念与作用。

2. 理解 Spring AOP 基于代理和代理工厂的核心工作原理，理解 Spring AOP 的主要实现过程。

3. 掌握使用 Spring 提供的代理工厂 ProxyFactoryBean 来实现 AOP 的方法，理解 Spring AOP 与 IoC 容器的结合方式以及各自起到的作用。

4. 掌握 Spring AOP 中 4 种类型的增强处理的作用和实现方法。

二、基本知识与原理

1. Spring AOP 的实现是基于代理和代理工厂的，由代理工厂创建代理对象，再由代理对象向目标对象的业务逻辑进行功能增强。

2. Spring 框架中提供了 ProxyFactory 和 ProxyFactoryBean 等多种代理工厂。使用 ProxyFactory，能够独立于 Spring 的 IoC 容器之外来使用 Spring 的 AOP 支持；而使用 ProxyFactoryBean，能使得 Spring AOP 与 IoC 容器有机地结合在一起，Spring 的 IoC 容器可以管理切点(Pointcut)和增强处理(Advice)等 AOP 组件。

3. Spring AOP 可以创建两种类型的代理对象：Java 动态代理和 CGLIB 代理。Java 动态代理只能针对接口生成代理对象(JdkDynamicAopProxy)，因此使用 Java 动态代理的对象必须实现至少一个接口；而 CGLIB 代理可以生成类级别的代理对象(Cglib2AopProxy)。

4. Spring AOP 中有 4 种增强处理的类型，如表 10-1 所示。

表 10-1 Spring AOP 中增强处理的类型

增强处理类型	接口	描述
Around	Org. aopalliance. intercept. MethodInterceptor	在目标方法调用前后切入
Before	Org. springframework. aop. MethodBeforeAdvice	在目标方法调用之前切入
After	Org. springframework. aop. AfterReturningAdvice	在目标方法调用之后切入
Throws	Org. springframework. aop. ThrowsAdvice	在目标方法抛出异常时切入

三、实验内容及步骤

1. 在 spring-prj3 中新建 cn. edu. zjut. advice 包，并创建 Before 类型的增强处理 LoggingAdvice.java，用于实现日志管理。其中的 before 方法定义了在目标方法调用之前需要切入的增强功能。具体代码如下所示。

```
package cn.edu.zjut.advice;

import org.springframework.aop.MethodBeforeAdvice;
import java.lang.reflect.Method;
```

```java
import java.util.Date;

public class LoggingAdvice implements MethodBeforeAdvice {

    public void before(Method m, Object[] args, Object target)
            throws Throwable {
        String date1 = (new Date()).toLocaleString();
        System.out.println("信息:[logging before][" + date1 + "]用户 "
                        + args[0] + " 尝试修改个人信息…");
    }
}
```

2. 修改 Spring 配置文件 applicationContext.xml,在其中增加对 LoggingAdvice 实例的配置,代码片段如下所示。

```xml
<!-- 增强处理定义 -->
<bean id="logAdvice" class="cn.edu.zjut.advice.LoggingAdvice" />
```

3. 修改 Spring 配置文件 applicationContext.xml,在其中通过 ProxyFactoryBean 对目标对象和增强处理等进行装配,代码片段如下所示。

```xml
<!-- 设定代理 -->
<bean id="logProxy"
    class="org.springframework.aop.framework.ProxyFactoryBean">

    <!-- 设定代理的是接口 -->
    <property name="proxyInterfaces">
        <value>cn.edu.zjut.service.IUserService</value>
    </property>

    <!-- 设定代理的目标类 -->
    <property name="target">
        <ref bean="userService"/>
    </property>

    <!-- 设定切入的 Advice -->
    <property name="interceptorNames">
        <list>
            <value>logAdvice</value>
        </list>
    </property>
</bean>
```

4. 修改 cn.edu.zjut.app 包中的测试类 SpringEnvTest,通过代理工厂返回 UserService 实例,并访问其 addUser()方法,具体代码如下所示。

```java
public class SpringEnvTest {
    public static void main(String[] args) {
        ApplicationContext ctx = new ClassPathXmlApplicationContext(
                "applicationContext.xml");
```

```
        IUserService userService = (IUserService)
            ctx.getBean("logProxy");
        ⋮
        userService.addUser(cust);
    }
}
```

5. 运行测试类 SpringEnvTest,观察控制台的输出,并记录运行结果。

6. 修改 cn.edu.zjut.advice 包中的 LoggingAdvice.java,使其同时实现 AfterReturningAdvice 接口,并添加 afterReturning 方法,用于定义在目标方法调用之后需要切入的增强功能、代码片段如下所示。

```
package cn.edu.zjut.advice;
import org.springframework.aop.AfterReturningAdvice;
⋮

public class LoggingAdvice implements MethodBeforeAdvice,
                    AfterReturningAdvice {

    public void afterReturning(Object returnValue, Method method,
            Object[] args, Object target) throws Throwable {
        String date2 = (new Date()).toLocaleString();
        System.out.println("信息:[logging after ][" + date2 + "]用户 "
                    + args[0] + " 成功修改个人信息…");

    }
    ⋮
}
```

7. 运行测试类 SpringEnvTest,观察控制台的输出,并记录运行结果。

8. 在 cn.edu.zjut.advice 包中创建 Around 类型的增强处理 LoggingInterceptor.java,并在其 invoke 方法中定义日志管理功能。其中的代码行"Object returnObject = invo.proceed();"不能缺少,需要通过它来调用目标对象的方法。具体代码如下所示。

```
package cn.edu.zjut.advice;

import java.util.Date;
import org.aopalliance.intercept.MethodInterceptor;
import org.aopalliance.intercept.MethodInvocation;

public class LoggingInterceptor implements MethodInterceptor {

    public Object invoke(MethodInvocation invo) throws Throwable {
        Object[] args = invo.getArguments();

        String date1 = (new Date()).toLocaleString();
        System.out.println("Interceptor 信息:[logging before ][" + date1
                + "]用户 " + args[0] + " 尝试修改个人信息…");

        Object returnObject = invo.proceed();
```

```
                String date2 = (new Date()).toLocaleString();
                System.out.println("Interceptor 信息:[logging after ][" + date2
                        + "]用户 " + args[0] + " 成功修改个人信息…");

            return args;
        }
}
```

9. 修改 Spring 配置文件 applicationContext.xml,在其中增加对 LoggingInterceptor 实例的配置,代码片段如下所示。

```
<!-- 增强处理定义 -->
<bean id = "logInterceptor"
        class = "cn.edu.zjut.advice.LoggingInterceptor" />
```

10. 修改 Spring 配置文件 applicationContext.xml,在代理设定中添加对 LoggingInterceptor 的装配,代码片段如下所示。

```
<bean id = "logProxy"
        class = "org.springframework.aop.framework.ProxyFactoryBean">

        <property name = "interceptorNames">
            <list>
                <value>logAdvice</value>
                <value>logInterceptor</value>
            </list>
        </property>
</bean>
```

11. 运行测试类 SpringEnvTest,观察控制台的输出,并记录运行结果。

四、实验要求

1. 填写并上交实验报告,报告中应包括如下内容。
(1) 运行结果截图。
(2) 根据实验过程,查找相关资料,总结使用 ProxyFactoryBean 实现 AOP 的基本步骤和主要配置方法,并记录下来。
(3) 根据实验过程,查找相关资料,总结 Spring AOP 中 4 种类型增强处理的作用及实现方法,并记录下来。
(4) 查找相关资料,思考如果需要代理的目标对象很多,可以借助于 Spring API 中的哪些接口或类来完成配置,并将其记录下来。
(5) 碰到的问题及解决方案或对问题的思考。
(6) 实验收获及总结。
2. 上交程序源代码,代码中应有相关注释。

第部分 企业级EJB组件编程技术

- 实验十一 会话Bean——用会话Bean实现用户登录及购物车应用
- 实验十二 实体Bean——用实体Bean实现用户信息的持久化
- 实验十三 消息驱动Bean——登录用户支付消息的分发应用

实验十一

会话 Bean——用会话 Bean 实现用户登录及购物车应用

基础实验——无状态会话 Bean 的调用

一、实验目的

1. 掌握 EJB 的概念。
2. 掌握 JBoss 服务器的安装与配置方法。
3. 掌握 JNDI 服务的发布过程。
4. 掌握会话 Bean 的开发步骤。

二、基本知识与原理

1. EJB(Enterprise Java Bean)是 sun 的 Java EE 服务器端组件模型,它定义了一个用于开发基于组件的企业多重应用程序的标准。凭借 Java 跨平台的优势,用 EJB 技术部署的分布式系统可以不限于特定的平台。

2. JNDI(Java Naming and Directory Interface,Java 命名和目录接口)是一组在 Java 应用中访问命名和目录服务的 API。命名服务将名称和对象联系起来,使得可用名称来访问对象。

3. 会话 Bean(Session Bean)用于执行业务流程的逻辑,属于客户端程序在服务器上的部分逻辑延伸。

三、实验内容及步骤

1. 下载 JBoss。登录 JBoss 官网 http://jbossas.jboss.org/downloads,选择 7.1.1 Final 版本,单击下载 JBoss AS(Application Server)服务器,如图 11-1 所示。

2. 安装运行 JBoss。解压下载成功的压缩包,然后打开 CMD 命令提示符窗口,进入 JBoss 解压主目录下的子目录 bin(如 C:\jboss-as-7.1.1.Final\bin)。运行 standlone.bat 来启动 JBoss 服务器(注意:要预先正确设置 JAVA_HOME 环境变量)。在 JBoss 服务器启动完毕之后,打开浏览器并输入地址 http://localhost:8080,如出现 JBoss 欢迎页面(如图 11-2 所示),则表示 JBoss 安装、运行成功。

3. 查看 Eclipse 的版本号。打开 Eclipse,选择 Help|About Eclipse 命令,在弹出的版

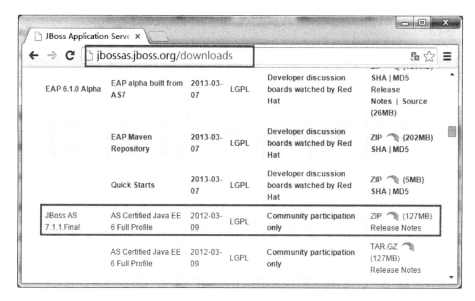

图 11-1　下载并解压安装 JBoss AS 7.1.1

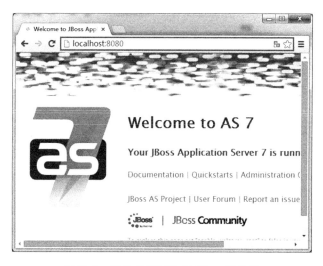

图 11-2　启动并测试访问 JBoss 服务器

本信息对话框中查看 Eclipse 的版本号,如图 11-3 所示(不同 Eclipse 的版本号可能不同,本例中 Eclipse 的版本号为 Kepler)。

4. 在 Eclipse 中安装 JBoss Tools 开发插件。

(1) 打开 Eclipse,选择 Help|Eclipse Marketplace 命令,如图 11-4 所示。

(2) 在弹出的对话框中的搜索编辑框中输入 JBoss Tools,然后在结果列表中找到 JBoss Tools(Kepler),单击 Install 按钮,如图 11-5 所示(注意:此操作需要连接到 Internet,并且所选择的 JBoss Tools 的版本号必须和 Eclipse 的版本号一致,如本例中二者版本号同为 Kepler)。

实验十一　会话Bean——用会话Bean实现用户登录及购物车应用

图 11-3　查看 Eclipse 的版本号

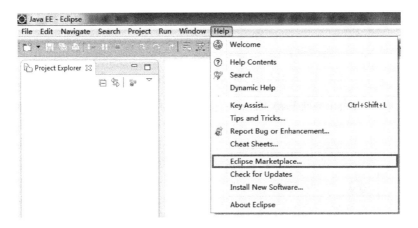

图 11-4　在 Eclipse 中查找 JBoss 开发插件

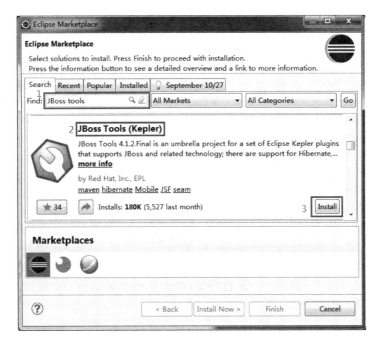

图 11-5　在 Eclipse 中安装 JBoss 开发插件

（3）在弹出的对话框中勾选 JBoss AS Tools 复选框，单击 Confirm 按钮，接着在安装许可协议对话框中选择 I accept the terms of the license agreement，如图 11-6 所示。

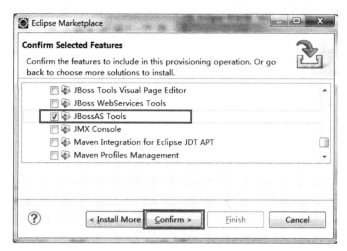

图 11-6　在 Eclipse Marketplace 中选择 JBoss AS 工具

（4）然后单击 Finish 按钮等待 Eclipse 自动下载并安装 JBossAS 开发插件，如图 11-6 和图 11-7 所示。

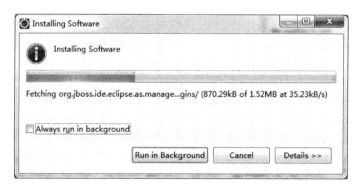

图 11-7　Eclipse 正在下载安装 JBossAS Tools 开发插件

（5）如在安装过程出现 Warning 警告，请单击 OK 按钮忽略警告。安装完毕后，重启 Eclipse 以使 JBossAS 开发插件生效。

5. 在 Eclipse 中配置 JBoss 服务器。

（1）打开 Eclipse，选择 Windows|Preferences 命令，在弹出的对话框的左侧导航栏中单击选择 Server|Runtime Environments，然后在右侧单击 Add 按钮，如图 11-8 所示。

（2）接着在弹出的服务器类型对话框中选择 JBoss Community|JBoss 7.1 Runtime 列表项，选中 Create a new local server 复选框，最后单击 Next 按钮，如图 11-9 所示。

（3）在弹出的 JBoss 服务器配置对话框中单击 Browse 按钮，选择 JBoss 的安装主目录（如 C:\jboss-as-7.1.1.Final）和 jre 运行环境，单击 Finish 按钮和 OK 按钮完成 JBoss 服务器配置，如图 11-10 和图 11-11 所示。

实验十一 会话 Bean —— 用会话 Bean 实现用户登录及购物车应用

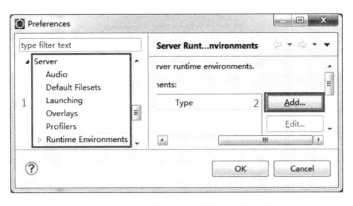

图 11-8 在 Eclipse 首选项对话框中添加新服务器

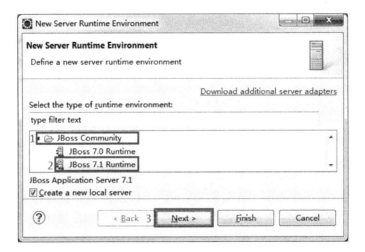

图 11-9 在 Eclipse 配置 JBoss 服务器之步骤 1

图 11-10 在 Eclipse 中配置 JBoss 服务器之步骤 2

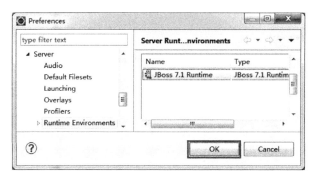

图11-11 完成JBoss服务器配置

6. 编写EJB工程(名称为ejb-project1)。

(1) 打开Eclipse,选择File|New|EJB Project命令,如图11-12所示。

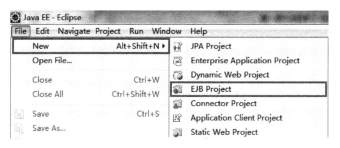

图11-12 创建EJB工程之步骤1

(2) 在弹出的对话框中填写工程名称ejb-project1,运行服务器选择上述步骤所配置的JBoss7.1 Runtime,EJB模块版本号选择3.0,然后单击Finish按钮完成工程的创建,如图11-13所示。

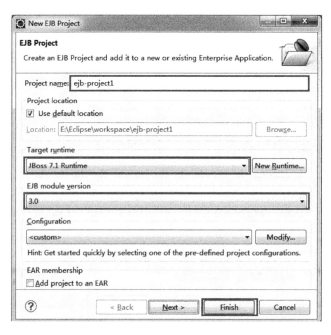

图11-13 创建EJB工程之步骤2

（3）在工程中添加一个 cn.edu.zjut.ejb 包。右击 ejb-project1 工程，在弹出的菜单中选择 New|Package，然后在弹出的对话框中填写包名称 cn.edu.zjut.ejb，单击 Finish 按钮。

（4）创建 UserService 会话 Bean。右击 cn.edu.zjut.ejb 包，在弹出的菜单中选择 New|Session Bean，然后在弹出的对话框中输入类名称 UserService，状态类型 Stateless，选中 Remote 复选框，取消选中 Local 复选框，最后单击 Finish 按钮，如图 11-14 所示。

图 11-14　填写会话 Bean 信息

（5）双击打开 UserServiceRemote.java，在其中添加一个抽象方法 login(String username，String password)，用于登录验证，具体代码如下所示。

```
package cn.edu.zjut.ejb;

import javax.ejb.Remote;

@Remote
public interface UserServiceRemote {
    public boolean login(String username, String password);
}
```

（6）双击打开 UserService.java，实现 UserServiceRemote 接口的 login(String username，String password)方法，具体代码如下所示。

```
package cn.edu.zjut.ejb;

import javax.ejb.Stateless;

@Stateless
public class UserService implements UserServiceRemote {
    public UserService() { }
    public boolean login(String username, String password){
```

```
        if(username.equals("zjut")&&password.equals("zjut")){
            return true;
        }else
            return false;
    }
}
```

(7) 将 ejb-project1 工程部署到 JBoss 服务器。右击 ejb-project1 工程,在弹出的菜单中选择 Export|EJB JAR file 命令,在弹出的对话框中单击 Browse 按钮,并在目录浏览对话框中选择 JBoss 主目录下的 standalone\deployments 子目录为部署地址 Destination(如: C:\jboss-as-7.1.1.Final\standalone\deployments),如图 11-15 所示。

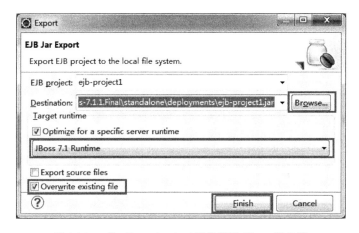

图 11-15　将 ejb-project1 工程部署到 JBoss 服务器

(8) 在 Eclipse 中启动 JBoss 服务器。选择 Eclipse 右下区域的 Servers 选项卡,在其中可以看到已经部署好的 JBoss 7.1 Runtime Server。单击该选项卡右侧的图标 ，启动 JBoss 服务器,如图 11-16 所示。

图 11-16　在 Eclipse 中启动 JBoss 服务器

(9) 服务器启动完成后,选择 Eclipse 右下区域的 Consoles 选项卡,查看 ejb-project1 工程已经部署到服务器的信息。在其中可以看到 UserService 服务发布地址为 module/UserService,如图 11-17 所示。

7. 编写客户端工程(名称为 ejb-project1-client)。

(1) 新建一个名称为 ejb-project1-client 的 Java Project 工程。

(2) 在工程的 src 目录下新建一个名称为 cn.edu.zjut.ejb 的包,将 EJB 工程 ejb-project1 中的接口程序 UserServiceRemote.java 复制到该包下,并添加类库 jboss-client.jar 到本工程的 Java Build Path-Libraries 中。

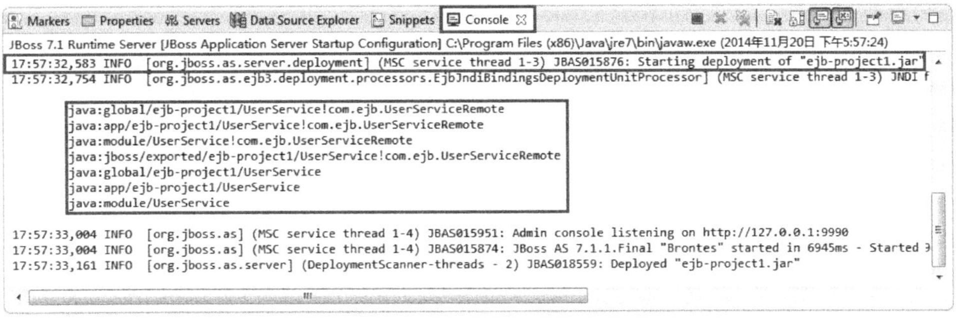

实验十一 会话 Bean——用会话 Bean 实现用户登录及购物车应用

图 11-17 ejb-project1 工程在 JBoss 服务器上部署成功的信息

(3) 在 src 目录下新建一个名称为 cn.edu.zjut.ejb.client 的包，在该包下新建一个名称为 LoginClient.java 的客户端程序，具体代码如下所示。

```java
package cn.edu.zjut.ejb.client;

import javax.naming.Context;
import javax.naming.InitialContext;
import javax.naming.NamingException;

import java.security.Security;
import java.util.Hashtable;

import cn.edu.zjut.ejb.UserServiceRemote;

public class LoginClient {
    private static UserServiceRemote lookupRemoteStatelessEjbBean() throws NamingException {
        final Hashtable jndiProperties = new Hashtable();
        jndiProperties.put(Context.URL_PKG_PREFIXES, "org.jboss.ejb.client.naming");
        final Context context = new InitialContext(jndiProperties);
        final String appName = "";
        final String moduleName = "ejb-project1";
        final String distinctName = "";
        final String beanName = "UserService";
        final String viewClassName = UserServiceRemote.class.getName();
        final String namespace = "ejb:" + appName + "/" + moduleName
                + "/" + distinctName + "/" + beanName + "!" + viewClassName;
        return (UserServiceRemote) context.lookup(namespace);
    }
    public static void main(String[] args) {
        //TODO Auto-generated method stub
        try{
            UserServiceRemote usBean = lookupRemoteStatelessEjbBean();
            System.out.println(usBean);
            boolean b1 = usBean.login("zjut","zjut");
            System.out.println(b1);
        }catch(NamingException e){
            e.printStackTrace();
        }
```

        }
    }

(4) 在 src 目录下新建 jboss-ejb-client.properties 文件,具体代码如下所示。

```
endpoint.name = client - endpoint
remote.connectionprovider.create.options.org.xnio.Options.SSL_ENABLED = false
remote.connections = default
remote.connection.default.host = 127.0.0.1
remote.connection.default.port  =  4447
remote.connection.default.connect.options.org.xnio.Options.SASL_POLICY_NOANONYMOUS = false
```

(5) 启动 JBoss7.1.1Final 服务器,然后运行 LoginClient.java,观察输出结果。
(6) 修改登录用户名和密码,观察输出结果。
(7) 修改 ejb-project1 工程的用户登录逻辑并重新部署。运行 LoginClient.java,观察输出结果。

四、实验要求

1. 填写并上交实验报告,报告中应包括如下内容。
(1) 运行结果截图。
(2) 修改后的关键代码及相应的运行结果或报错信息。
(3) 根据实验过程,总结客户端程序调用 EJB 的过程。
(4) 碰到的问题及解决方案或对问题的思考。
(5) 实验收获及总结。
2. 上交程序源代码,代码中应有相关注释。

提高实验——有状态会话 Bean 的调用

一、实验目的

1. 掌握有状态会话 Bean 的编写与调用方法。
2. 掌握在 Web 工程中调用有状态会话 Bean 和无状态会话 Bean 的方法。

二、基本知识与原理

1. 有状态会话 Bean 和无状态会话 Bean 都实现了 javax.ejb.SessionBean 接口。有状态会话 Bean 可以在多次访问之间保存特定用户的信息,而无状态会话 Bean 不会在用户多次访问之间保存信息。
2. 无状态会话 Bean 在对象实例池中进行创建和维护,可以提供给众多用户共同使用;有状态会话 Bean 在一个生命周期内只服务于一个用户。

三、实验内容及步骤

1. 编写 Web 客户端工程(名称为 ejb-project1-web)。
(1) 新建一个名称为 ejb-project1-web 的 Dynamic Web Project 工程。

实验十一　会话 Bean —— 用会话 Bean 实现用户登录及购物车应用

（2）在 src 目录下新建一个名称为 cn.edu.zjut.ejb 的包，将 EJB 工程 ejb-project1 中的接口程序 UserServiceRemote.java 复制到该包下。

（3）在 src 目录下新建 jboss-ejb-client.properties 文件（代码略），并添加类库 jboss-client.jar 到本工程的 Java Build Path-Libraries 中。

（4）在 WebContent 目录下新建一个名称为 webclient.jsp 的 JSP 页面，并在其中的 WEB-INF 目录下新建 web.xml 文件。整个工程目录如图 11-18 所示。

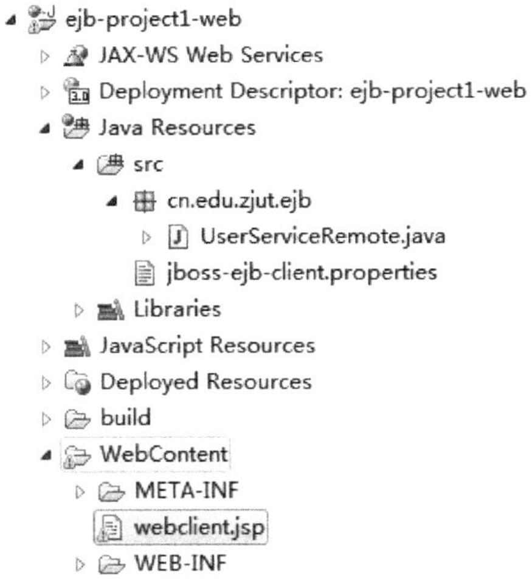

图 11-18　ejb-project1-web 工程结构

（5）Webclient.jsp 页面的具体代码如下所示。

```jsp
<%@ page language="java" import="java.util.*" pageEncoding="UTF-8"%>
<%@ page import="javax.naming.*,java.util.Properties"%>
<%@ page import="cn.edu.zjut.ejb.*"%>

<%
    try{
        final Hashtable jndiProperties = new Hashtable();
        jndiProperties.put(Context.URL_PKG_PREFIXES, "org.jboss.ejb.client.naming");
        final Context context = new InitialContext(jndiProperties);
        final String appName = "";
        final String moduleName = "ejb-project1";
        final String distinctName = "";
        final String beanName = "UserService";
        final String viewClassName = UserServiceRemote.class.getName();
        final String namespace = "ejb:" + appName + "/" + moduleName
            + "/" + distinctName + "/" + beanName + "!" + viewClassName;
        UserServiceRemote usBean = (UserServiceRemote) context.lookup(namespace);
        System.out.println(usBean);
        if(usBean.login("zjut","zjut"))
            out.println("login ok!");
        else
            out.println("login failed!");
    }catch(NamingException e){
```

```
            e.printStackTrace();
    }
%>
```

（6）将工程部署到 JBoss 服务器上，然后打开浏览器输入网址 http://localhost:8080/ejb-project1-web/webclient.jsp，查看运行结果。

（7）改变 ejb-project1 工程中的用户登录逻辑，然后将工程重新部署到 JBoss 服务器上，再次观察 webclient.jsp 的运行结果。

2. 编写有状态会话 Bean。

（1）修改 ejb-project1 工程，在 cn.edu.zjut.ejb 包中新建一个名称为 ProductCartBean 的有状态会话 Bean，选择 State Type 下拉列表框中的 Stateful 选项，如图 11-19 所示。

图 11-19　编写有状态会话 Bean

（2）编写 ProductCartRemote.java 代码如下所示。

```
package cn.edu.zjut.ejb;
import java.util.ArrayList;
import javax.ejb.Remote;

@Remote
public interface ProductCartRemote {
    public void addProduct(String productName, int price);
    public ArrayList<String> listProducts();
    public int totalPrice();
}
```

（3）编写 ProductCartBean.java 代码如下所示。

```java
package cn.edu.zjut.ejb;
import java.util.ArrayList;
import javax.ejb.Stateful;

@Stateful
@Remote(ProductCartRemote.class)
public class ProductCartBean implements ProductCartRemote {
    public ProductCartBean() {
        //TODO Auto-generated constructor stub
    }
    private ArrayList<String> cartList = new ArrayList<String>();
    private int totalPrice = 0;
    public ArrayList listProducts(){
        return this.cartList;
    }
    public void addProduct(String name, int price){
        this.cartList.add(name);
        totalPrice += price;
    }
    public int totalPrice(){return totalPrice;}
}
```

（4）将新修改的有状态会话 Bean 部署到 JBoss 服务器。

3. 修改 ejb-project1-web 工程。

（1）将 EJB 工程 ejb-project1 中的接口程序 ProductCartRemote.java 复制到 cn.edu.zjut.ejb 包下。

（2）在 WebContent 目录下新建一个名称为 myCart.jsp 的 JSP 页面，具体代码如下所示。

```jsp
<%@ page language="java" import="java.util.*" pageEncoding="UTF-8"%>
<%@ page import="javax.naming.*, java.util.Properties"%>
<%@ page import="cn.edu.zjut.ejb.*"%>

<%
    try{
        final Hashtable jndiProperties = new Hashtable();
        jndiProperties.put(Context.URL_PKG_PREFIXES, "org.jboss.ejb.client.naming");
        final Context context = new InitialContext(jndiProperties);
        final String appName = "";
        final String moduleName = "ejb-project1";
        final String distinctName = "";
        final String beanName = "ProductCartBean";
        final String viewClassName = ProductCartRemote.class.getName();
        final String namespace = "ejb:" + appName + "/" + moduleName
            + "/" + distinctName + "/" + beanName + "!" + viewClassName
            + "?stateful";
        ProductCartRemote cart = null;
        cart = (ProductCartRemote)session.getAttribute("cart");
        if(cart == null){
            cart = (ProductCartRemote) context.lookup(namespace);
```

```jsp
            session.setAttribute("cart",cart);
        }else{
            String productName = request.getParameter("product");
            String sPrice = request.getParameter("price");
            int price = 0;
            if(sPrice!= null) price = Integer.parseInt(sPrice);
            cart.addProduct(productName, price);
            List<String> myProducts = cart.listProducts();
            out.println("Total Price:" + cart.totalPrice() + "<br>");
            out.println("My Products:<br>" + myProducts);
        }
    }catch(NamingException e){
        e.printStackTrace();
    }
%>
<table border = 1>
  <tr><td><a href = "myCart.jsp?product = fridge&price = 3000"> fridge buy </a></td></tr>
  <tr><td><a href = "myCart.jsp?product = ledtv&price = 5000"> ledtv buy </a></td></tr>
  <tr><td><a href = "myCart.jsp?product = waterheater&price = 2800"> waterheater buy </a></td></tr>
  <tr><td><a href = "myCart.jsp?product = car&price = 300000"> car buy </a></td></tr>
</table>
```

（3）将工程部署到 JBoss 服务器上,然后打开浏览器,输入网址 http://localhost:8080/ejb-project1-web/myCart.jsp,单击页面中的"购买"按钮并查看运行结果。

四、实验要求

1. 填写并上交实验报告,报告中应包括如下内容:
（1）实验基本思路。
（2）实验关键代码及相应的运行结果及截图,或相应的报错信息。
（3）碰到的问题及解决方案或对问题的思考。
（4）实验收获及总结。
2. 上交程序源代码,代码中应有相关注释。

扩展实验——控制会话 Bean 的生命周期

一、实验目的

1. 掌握会话 Bean 生命周期的概念。
2. 掌握 EJB 的关于生命周期的 6 个注解符:@postConstruct、@PreDestroy、@PrePassivate、@PostActivate、@Init 和@Remove。

二、基本知识与原理

1. 会话 Bean 实例的生命周期一般较短,包括 Bean 对象的创建、初始化、运行、钝化、激活、销毁。
2. 无状态会话 Bean 对应生命周期管理的注解符包括@postConstruct 和@PreDestroy。

3. 有状态会话 Bean 对应生命周期管理的注解符包括@postConstruct、@PreDestroy、@PrePassive、@PostActivate、@Init 和@Remove。

4. 各注解符含义如下。

@postConstruct：当 bean 对象完成实例化后，标注了这个注解符的方法会立即被调用，每个 bean 类只能定义一个@PostConstruct 方法。

@PreDestroy：标注了这个注解符的方法会在容器销毁一个无用或者过期的 bean 实例之前被调用。

@ PrePassivate：当一个有状态的 bean 实例空闲时间过长，就会发生钝化（passivate）。标注了该注解符的方法会在被钝化之前被调用，bean 实例被钝化后，在一段时间内，如果仍然没有用户对 bean 实例进行操作，容器将会从硬盘中删除它，这个注释符用于有状态会话 bean。

@PostActivate：当客户端再次使用已经被钝化的有状态 bean 时，EJB 容器会重新实例化一个 Bean 实例，并从硬盘中将之前的状态恢复。标注了这个注解符的方法会在激活完成时被调用。这个注释符只适合于有状态会话 bean。

@Init：这个注释符指定了有状态 bean 初始化的方法，它区别于@PostConstruct 注释之处在于：多个@Init 注释方法可以同时存在于有状态 session bean 中，但每个 bean 实例只会有一个@Init 注释方法被调用，@PostConstruct 在@Init 之后调用。

@Remove：当客户端调用标注了@Remove 注解符的方法时，容器将在方法执行结束后把 bean 实例删除。

三、实验内容及步骤

1. 修改 ejb-project1 工程，新建一个名称为 cn.edu.zjut.ejb.lifecycle 的包。

2. 在 cn.edu.zjut.ejb.lifecycle 包下新建一个名称为 LifeCycle 的有状态会话 Bean，如图 11-20 所示。

图 11-20　新建一个有状态会话 Bean

3. 为 LifeCycle 类添加一个接口方法 hello 及 6 个注释符的方法,具体代码如下:

//LifeCycleRemote.java

```java
package cn.edu.zjut.ejb.lifecycle;

import javax.ejb.Remote;

@Remote
public interface LifeCycleRemote {
    public String hello(String name);
    public void removeSession();
}
```

//LifeCycle.java

```java
package cn.edu.zjut.ejb.lifecycle;

import javax.annotation.PostConstruct;
import javax.annotation.PreDestroy;
import javax.ejb.Init;
import javax.ejb.PostActivate;
import javax.ejb.PrePassivate;
import javax.ejb.Remove;
import javax.ejb.Stateful;

@Stateful
public class LifeCycle implements LifeCycleRemote {
    public LifeCycle() {
        //TODO Auto-generated constructor stub
    }
    public String hello(String name){
        try {
            Thread.sleep(1000 * 3);
        } catch(InterruptedException e) {
            e.printStackTrace();
        }
        return "hello" + name + "!";
    }

    @Init
    public void initialize() {
        System.out.println("@Init 事件触发");
    }

    @PostConstruct
    public void construct() {
        System.out.println("@PostConstruct 事件触发");
    }

    @PreDestroy
```

```java
    public void exit() {
        System.out.println("@PreDestroy事件触发");
    }

    @PostActivate
    public void activate() {
        System.out.println("@PostActivate事件触发");
    }

    @PrePassivate
    public void prePassivate() {
        System.out.println("@prePassivate事件触发");
    }

    @Remove
    public void removeSession() {
        System.out.println("@remove事件触发");
    }
}
```

4. 将 ejb-project1 工程部署到 JBoss 服务器。

5. 修改客户端工程 ejb-project1-client：

（1）在工程中新建一个名称为 cn.edu.zjut.ejb.lifecycle 的包。

（2）将 EJB 工程 ejb-project1 中的接口程序 LifeCycleRemote.java 和 LifeCycle.java 复制到本工程的 cn.edu.zjut.ejb.lifecycle 包下。

（3）在 cn.edu.zjut.ejb.client 包下新建一个名称为 LifeCycleClient.java 的客户端程序，具体代码如下：

```java
package cn.edu.zjut.ejb.client;

import javax.naming.Context;
import javax.naming.InitialContext;
import javax.naming.NamingException;

import java.security.Security;
import java.util.Hashtable;

import cn.edu.zjut.ejb.lifecycle.*;

public class LifeCycleClient {
    private static LifeCycleRemote lookupRemoteStatefulEjbBean() throws NamingException {

        final Hashtable jndiProperties = new Hashtable();
        jndiProperties.put(Context.URL_PKG_PREFIXES, "org.jboss.ejb.client.naming");

        final Context context = new InitialContext(jndiProperties);
        final String appName = "";

        final String moduleName = "ejb-project1";
```

```java
            final String distinctName = "";

            final String beanName = LifeCycle.class.getSimpleName();

            final String viewClassName = LifeCycleRemote.class.getName();

            final String namespace = "ejb:" + appName + "/" + moduleName
                + "/" + distinctName + "/" + beanName + "!" + viewClassName + "?stateful";

            System.out.println("namespace = " + namespace);

            return (LifeCycleRemote) context.lookup(namespace);
    }

    public static void main(String[] args) {
        //TODO Auto-generated method stub
        try{
            LifeCycleRemote lc = lookupRemoteStatefulEjbBean();
            lc.hello("zjut");
            lc.removeSession();
        }catch(NamingException e){
            e.printStackTrace();
        }
    }
}
```

6. 运行 LifeCycleClient.java 程序,查看结果。

四、实验要求

1. 填写并上交实验报告,报告中应包括如下内容:
(1) 实验基本思路;
(2) 实验关键代码、相应的运行结果及截图,或相应的报错信息;
(3) 碰到的问题及思考;
(4) 实验收获及总结。
2. 上交程序源代码,代码中应有相关注释。

实验十二

实体 Bean——用实体 Bean 实现用户信息的持久化

基础实验——实体 Bean 的开发

一、实验目的

1. 掌握实体 Bean 的概念和 JPA 规范。
2. 掌握实体 Bean 的开发步骤。
3. 掌握通过实体管理器来执行数据库更新的方法。
4. 掌握实体 Bean 的监听和回调方法。
5. 掌握使用 JPQL 查询语言执行数据库实体查询的方法。

二、基本知识与原理

1. 实体 Bean(Entity Bean)是持久数据组件,代表存储在外部介质中的持久(Persistence)对象或者已有的企业应用系统资源。一个实体 Bean 可以代表数据库中的一行记录,多个客户端应用能够以共享方式访问表示该数据库记录的实体 Bean。

2. JPA(Java Persistence API,Java 持久化接口)是指通过注解或 XML 来描述对象到关系表的映射,并将运行期的实体对象持久化到数据库中。通过使用 JPA,开发人员不再局限于私有供应商提供的特有 API,除非该功能是供应商特有的。

3. JPQL(Java Persistence Query Language,Java 持久化查询语言)是一种可移植的查询语言,旨在以面向对象的表达式语言将 SQL 语法和简单查询语义绑定在一起。使用这种语言编写的查询是可移植的,可以被编译成所有主流数据库服务器上的 SQL。JPQL 是完全面向对象的,具备继承、多态和关联等特性。

三、实验内容及步骤

1. 下载并安装 MySQL5.5 版本。

(1) 输入 http://dev.mysql.com/downloads/mysql/5.5.html#downloads,选择适合当前系统的 MySQL 版本(如 x86,32-bit,5.5.41),单击 Download 按钮,如图 12-1 所示。

(2) 在弹出的页面输入账号和密码(如果没有可以立即注册一个),继续完成下载及安装任务。

```
Other Downloads:
Windows (x86, 32-bit), MSI Installer    5.5    39.2M    Download
```

图 12-1　下载 MySQL

（3）安装 MySQL 完毕后，按照系统提示完成实例配置（注意：务必记住为 root 用户设置的密码）。

（4）为了便于操作 MySQL 数据库，建议安装配套的 MySQL workbench 版本（例如 MySQL Workbench6.1）。

2. 创建数据库 EIS。

（1）使用 root 用户在 MySQL 中创建一个名称为 EIS 的数据库。

（2）在 EIS 数据库中创建一个名称为 userlist 的表，如表 12-1 所示。

表 12-1　userlist 数据表

字 段 名 称	类　　型	中　文　名
userid	int	用户 ID，自动增长
username	Varchar(50)	用户名
userpwd	Varchar(50)	密码
age	int	年龄

具体代码如下所示。

```
CREATE TABLE 'userlist'(
  'userid' int(11) AUTO_INCREMENT,
  'age' int(11) DEFAULT NULL,
  'username' varchar(50) DEFAULT NULL,
  'userpwd' varchar(50) DEFAULT NULL,
  PRIMARY KEY ('userid')
) ENGINE = InnoDB DEFAULT CHARSET = gb2312;
```

（3）为 userlist 表中添加 10 条数据。

3. 在 JBoss 中配置数据源。

（1）在 %JBOSS_HOME%\modules\com 目录下创建 mysql\main 目录（如 C:\jboss-as-7.1.1.Final\modules\com\mysql\main），将 MySQL 驱动程序复制到此目录下（如 mysql-connector-java-5.1.26-bin.jar），并创建 module.xml 文件，具体代码如下所示。

```
<?xml version = "1.0" encoding = "UTF-8"?>
<module xmlns = "urn:jboss:module:1.1" name = "com.mysql">
    <resources>
        <resource-root path = "mysql-connector-java-5.1.26-bin.jar"/>
    </resources>
    <dependencies>
        <module name = "javax.api"/>
        <module name = "javax.transaction.api"/>
        <module name = "javax.servlet.api" optional = "true"/>
    </dependencies>
</module>
```

实验十二 实体 Bean——用实体 Bean 实现用户信息的持久化

(2) 修改%JBOSS_HOME%\standalone\configuration 目录下的(如 C:\jboss-as-7.1.1.Final\standalone\configuration)XML 配置文件 standalone.xml 中的<datasources>元素，修改完毕后的元素内容如下(加粗字体为新增内容)所示。

```xml
<datasources>
    <datasource jndi-name="java:jboss/datasources/ExampleDS" pool-name="ExampleDS" enabled="true" use-java-context="true"><connection-url>jdbc:h2:mem:test;DB_CLOSE_DELAY=-1</connection-url>
        <driver>h2</driver>
        <security>
            <user-name>sa</user-name>
            <password>sa</password>
        </security>
    </datasource>
    <datasource jndi-name="java:/MySqlDS" pool-name="MySqlDS" enabled="true" use-java-context="true"><connection-url>jdbc:mysql://localhost:3306/eis</connection-url>
        <driver>mysql</driver>
        <pool>
            <min-pool-size>20</min-pool-size>
            <max-pool-size>20</max-pool-size>
            <prefill>true</prefill>
        </pool>
        <security>
            <user-name>root</user-name>
            <password>root</password>
        </security>
    </datasource>
    <drivers>
        <driver name="h2" module="com.h2database.h2">
            <xa-datasource-class>org.h2.jdbcx.JdbcDataSource</xa-datasource-class>
        </driver>
        <driver name="mysql" module="com.mysql">
            <driver-class>com.mysql.jdbc.Driver</driver-class>
            <xa-datasource-class>com.mysql.jdbc.jdbc2.optional.MysqlXADataSource</xa-datasource-class>
        </driver>
    </drivers>
</datasources>
```

(3) 进入%JBOSS_HOME%\bin 目录，运行 add-user.bat 文件，新建一个用户名为 appuser、密码为 apppwd 的普通用户(Application User)，以及一个用户名为 admin、密码为 manager 的管理用户(Management User)。

(4) 运行%JBOSS_HOME%\bin 目录下的 standalone.bat，启动 JBoss 服务器。

(5) 打开浏览器，输入网址 http://127.0.0.1:9990/console/App.html#ds-metrics，在弹出的登录窗口中输入用户名 admin 和密码 manager。登录后在左侧导航栏查看 Subsystems|Metrics|Datasources 选项卡，可以看到图 12-2 所示的 MySqlDS，这表示数据

源配置成功。或者，在控制台看到如下输出信息，这也表示数据源配置成功。

```
08:55:07,193 INFO
[org.jboss.as.connector.subsystems.datasources] (MSC service thread 1-1) JBAS010400: Bound
data source [java:/MySqlDS]
```

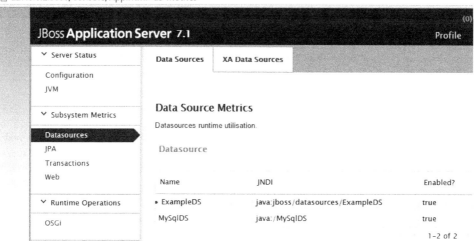

图 12-2　在管理后台查看数据源

4. 创建 EJB 工程。

（1）新建一个名称为 ejb-project2 的 EJB Project 工程，在其 Java Build Path|Libraries 中添加一个 External JAR 库文件 hibernate-entitymanager-4.0.1.Final.jar。该文件所在位置为%JBOSS_HOME%\modules\org\hibernate\main。

（2）在 ejbModule\META-INF 下创建一个名称为 persistence.xml 的持久化配置文件，具体代码如下所示。

```xml
<?xml version = "1.0" encoding = "UTF-8" ?>
<persistence xmlns = "http://java.sun.com/xml/ns/persistence"
    xmlns:xsi = "http://www.w3.org/2001/XMLSchema-instance"
    xsi:schemaLocation = "http://java.sun.com/xml/ns/persistence

http://java.sun.com/xml/ns/persistence/persistence_1_0.xsd" version = "1.0">
    <persistence-unit name = "userpu">

        <provider>org.hibernate.ejb.HibernatePersistence</provider>
        <jta-data-source>java:/MySqlDS</jta-data-source>
        <class>cn.edu.zjut.ejb.User</class>
        <properties>
            <property name = "hibernate.connection.autocommit" value = "false" />
            <property name = "hibernate.dialect" value = "org.hibernate.dialect.MySQL5Dialect" />
            <property name = "hibernate.show_sql" value = "true" />
            <property name = "hibernate.format_sql" value = "true" />
            <property name = "hibernate.hbm2ddl.auto" value = "update" />
        </properties>
```

实验十二 实体 Bean——用实体 Bean 实现用户信息的持久化

```
    </persistence-unit>
</persistence>
```

(3) 在 ejbModule 下创建 cn.edu.zjut.ejb 包,并在其中新建一个名称为 User.java 的实体 Bean,具体代码如下所示。

```java
package cn.edu.zjut.ejb;

import java.io.Serializable;
import javax.persistence.Column;
import javax.persistence.Entity;
import javax.persistence.GeneratedValue;
import javax.persistence.GenerationType;
import javax.persistence.Id;
import javax.persistence.Table;

@Entity
@Table(name = "userlist")

public class User implements Serializable {
    @Id
    @GeneratedValue(strategy = GenerationType.AUTO)
    private Integer userid;

    @Column(name = "username", length = 50)
    private String username;

    @Column(name = "userpwd", length = 50)
    private String userpwd;

    @Column(name = "age")
    private Integer age;

    public Integer getUserid() {
        return userid;
    }
    public void setUserid(Integer userid) {
        this.userid = userid;
    }

    public String getUsername() {
        return username;
    }
    public void setUsername(String username) {
        this.username = username;
    }

    public String getUserpwd() {
        return userpwd;
    }
    public void setUserpwd(String userpwd) {
```

```java
        this.userpwd = userpwd;
    }

    public Integer getAge() {
        return age;
    }

    public void setAge(Integer age) {
        this.age = age;
    }
}
```

（4）在 ejbModule 下创建 cn.edu.zjut.ejb.dao 包，并在其中新建名称分别为 UserDAO.java 和 UserDAORemote.java 的无状态会话 Bean，具体代码如下所示。

```java
//UserDAORemote.java
package cn.edu.zjut.ejb.dao;

import java.util.List;
import javax.ejb.Remote;
import cn.edu.zjut.ejb.User;

@Remote
public interface UserDAORemote {
    public List<User> select(String sql);
    public boolean insert(User user);
}

//UserDAO.java
package cn.edu.zjut.ejb.dao;

import java.util.List;

import javax.ejb.Stateless;
import javax.persistence.*;

import cn.edu.zjut.ejb.*;

@Stateless
public class UserDAO implements UserDAORemote {
    @PersistenceContext(type = PersistenceContextType.EXTENDED, unitName = "userpu")
    EntityManager em;
    public UserDAO() {
    }
    public boolean insert(User user){
        try{
            em.persist(user);
        }catch(Exception e){
            e.printStackTrace();
            return false;
        }
```

```
            return true;
    }
    public List < User > select(String sql){
        List < User > userlist = null;
        try{
            Query q = em.createQuery(sql);
            userlist = (List < User >)q.getResultList();
        }catch(Exception e){
            e.printStackTrace();
            return userlist;
        }
        return userlist;
    }
}
```

（5）整个工程目录结构如图 12-3 所示。

（6）右击 ejb-project2 工程，选择 Export|EJB Jar File 命令将工程部署到 JBoss 服务器。

5. 创建 EJB 客户端工程。

（1）新建一个名称为 ejb-project2-client 的 Java Project 工程，在 ejb-project2-client 的 Java Build Path | Libraries 中添加 External JAR 库文件 hibernate-jpa-2.0-api-1.0.1.Final.jar，该文件所在位置为%JBOSS_HOME%\modules\javax\persistence\api\main。添加 External JAR 库文件 jboss-client.jar，该文件所在位置为%JBOSS_HOME%\bin\client。

图 12-3　ejb-project2 工程目录结构

（2）在 src 目录下新建一个名称为 cn.edu.zjut.ejb 的包，将工程 ejb-project2 的 User.java 复制到其中。

（3）在 src 目录下新建一个名称为 cn.edu.zjut.ejb.dao 的包，将工程 ejb-project2 的 UserDAORemote.java 复制到其中。

（4）在 src 目录下新建一个名称为 jboss-ejb-client.properties 的文件，具体代码如下所示。

```
endpoint.name = client - endpoint
remote.connectionprovider.create.options.org.xnio.Options.SSL_ENABLED = false

remote.connections = default

remote.connection.default.host = 127.0.0.1
remote.connection.default.port = 4447
remote.connection.default.connect.options.org.xnio.Options.SASL_POLICY_NOANONYMOUS = false

remote.connection.default.username = appuser
remote.connection.default.password = apppwd
```

（5）在 src 目录下新建一个名称为 cn.edu.zjut.ejb.client 的包，新建一个文件 UserDAOTest

.java,具体代码如下所示。

```java
package cn.edu.zjut.ejb.client;

import java.util.Hashtable;
import java.util.List;

import javax.naming.Context;
import javax.naming.InitialContext;
import javax.naming.NamingException;

import cn.edu.zjut.ejb.dao.UserDAORemote;
import cn.edu.zjut.ejb.User;

public class UserDAOTest {
private static UserDAORemote lookupRemoteStatelessEjbBean() throws NamingException {

        final Hashtable jndiProperties = new Hashtable();
        jndiProperties.put(Context.URL_PKG_PREFIXES, "org.jboss.ejb.client.naming");
        final Context context = new InitialContext(jndiProperties);
        final String appName = "";
        final String moduleName = "ejb-project2";
        final String distinctName = "";
        final String beanName = "UserDAO";
        final String viewClassName = UserDAORemote.class.getName();
        final String namespace = "ejb:" + appName + "/" + moduleName
            + "/" + distinctName + "/" + beanName + "!" + viewClassName;
        System.out.println(namespace);
        return (UserDAORemote) context.lookup(namespace);
    }

    public static void main(String[] args) {
        try{
            UserDAORemote userdao = lookupRemoteStatelessEjbBean();
            System.out.println(userdao);
            List<User> userlist = userdao.select("select u from User u");
            System.out.println("userlist = " + userlist);
            User u = (User)userlist.get(0);
            System.out.println("username = " + u.getUsername());
            System.out.println("age = " + u.getAge());
        }catch(NamingException e){
            e.printStackTrace();
        }
    }
}
```

(6) 整个 ejb-project2-client 工程目录结构如图 12-4 所示。

(7) 启动 JBoss7.1.1Final 服务器,运行 UserDAOTest.java 文件,查看运行结果。

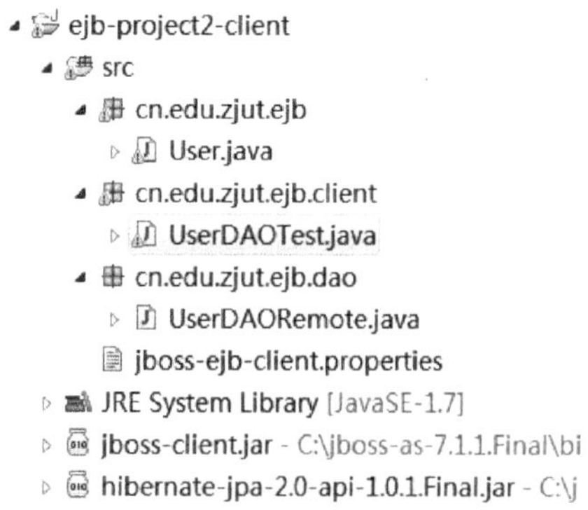

图 12-4　ejb-project2-client 工程目录结构

四、实验要求

1. 填写并上交实验报告，报告中应包括如下内容。
（1）运行结果截图。
（2）修改后的关键代码及相应的运行结果或报错信息。
（3）根据实验过程，总结实体 Bean 的具体开发过程。
（4）碰到的问题及解决方案或对问题的思考。
（5）实验收获及总结。
2. 上交程序源代码，代码中应有相关注释。

提高实验——使用 JPQL 语言

一、实验目的

1. 掌握 JPQL 查询语言（Java Persistence Query Language）的基本语法。
2. 掌握使用 JPQL 来添加、删除、查询、修改数据库的方法。

二、基本知识与原理

JPQL 是在 EJB2.0 中引入的一种可移植的查询语言，旨在以面向对象表达式语言的表达式，将 SQL 语法和简单查询语义绑定在一起。使用这种语言编写的查询是可移植的，可以被编译成所有主流数据库服务器上的 SQL。

三、实验内容及步骤

1. 打开 ejb-project2 工程，新建一个 cn.edu.zjut.ejb.dao.jpql 包，并在该包下新建名称分别为 JPQLUserDAORemote.java 和 JPQLUserDAO.java 的无状态会话 Bean。具体代码如下所示。

```
//JPQLUserDAORemote.java
package cn.edu.zjut.ejb.dao.jpql;

import java.util.List;
import javax.ejb.Remote;
import cn.edu.zjut.ejb.User;

@Remote
public interface JPQLUserDAORemote {
    public boolean insert(User user);
    public boolean update(User user);
    public boolean deleteById(int userid);
    public User selectById(int userid);
    public List<User> list(int pageSize, int pageNo);
}
```

```java
//JPQLUserDAO.java
package cn.edu.zjut.ejb.dao.jpql;

import java.util.List;

import javax.ejb.LocalBean;
import javax.ejb.Stateless;
import javax.persistence.EntityManager;
import javax.persistence.PersistenceContext;
import javax.persistence.PersistenceContextType;
import javax.persistence.Query;

import cn.edu.zjut.ejb.User;

@Stateless
@LocalBean
public class JPQLUserDAO implements JPQLUserDAORemote {
    @PersistenceContext(type = PersistenceContextType.EXTENDED, unitName = "userpu")
    EntityManager em;
    public JPQLUserDAO() {
    }

    public boolean insert(User user){
        try{
            em.persist(user);
        }catch(Exception e){
            e.printStackTrace();
            return false;
        }
        return true;
    }

    public boolean update(User user){
        try{
            em.merge(user);
        }catch(Exception e){
            e.printStackTrace();
            return false;
        }
        return true;
    }

    public boolean deleteById(int userid){
        try{
            Query q = em.createQuery("delete from User u where u.userid = ?1");
            q.setParameter(1, new Integer(userid));
            q.executeUpdate();
        }catch(Exception e){
            e.printStackTrace();
            return false;
        }
```

```java
            return true;
        }

        public User selectById(int userid){
            User user = null;
            try{
                Query q = em.createQuery("select u from User u where u.userid = ?1");
                q.setParameter(1, new Integer(userid));
                user = (User)q.getSingleResult();
            }catch(Exception e){
                e.printStackTrace();
                return user;
            }
            return user;
        }

        public List<User> list(int pageSize, int pageNo){
            List<User> userlist = null;
            try{
                int index = (pageNo - 1) * pageSize;
                Query q = em.createQuery("select u from User u order by userid asc");
    userlist = (List<User>)q.setMaxResults(pageSize).setFirstResult(index).getResultList();
                em.clear();
            }catch(Exception e){
                e.printStackTrace();
                return userlist;
            }
            return userlist;
        }
}
```

2. 打开 ejb-project2-client 工程，新建一个 cn.edu.zjut.ejb.dao.jpql 包，将 ejb-project2 工程中的 JPQLUserDAORemote.java 复制到其中。

3. 在 ejb-project2-client 工程的 cn.edu.zjut.ejb.client 包下新建名称为 JPQLUserDAOTest.java 的 Java 类，具体代码如下所示。

```java
//JPQLUserDAOTest.java
package cn.edu.zjut.ejb.client;

import java.util.Hashtable;
import java.util.List;

import javax.naming.Context;
import javax.naming.InitialContext;
import javax.naming.NamingException;

import cn.edu.zjut.ejb.dao.jpql.JPQLUserDAORemote;
import cn.edu.zjut.ejb.User;
```

```java
public class JPQLUserDAOTest {
    private static JPQLUserDAORemote lookupRemoteStatelessEjbBean() throws NamingException {

        final Hashtable jndiProperties = new Hashtable();
        jndiProperties.put(Context.URL_PKG_PREFIXES, "org.jboss.ejb.client.naming");

        final Context context = new InitialContext(jndiProperties);
        final String appName = "";

        final String moduleName = "ejb-project2";

        final String distinctName = "";

        final String beanName = "JPQLUserDAO";

        final String viewClassName = JPQLUserDAORemote.class.getName();

        final String namespace = "ejb:" + appName + "/" + moduleName
            + "/" + distinctName + "/" + beanName + "!" + viewClassName;

        System.out.println(namespace);

        return (JPQLUserDAORemote) context.lookup(namespace);

    }

    public static void main(String[] args) {
        //TODO Auto-generated method stub
        try{
            JPQLUserDAORemote userdao = lookupRemoteStatelessEjbBean();
            System.out.println(userdao);
            User user = new User();
            user.setAge(6);
            user.setUsername("new6");
            user.setUserpwd("new6");

            //插入 Insert
            if(userdao.insert(user))
                System.out.println("new record has been inserted!");
            else
                System.out.println("insert failed!");
            //查询 Select
            user = null;
            user = userdao.selectById(1);
            System.out.println("new username = " + user.getUsername());
            //更新 update
            user.setUsername("newname");
            userdao.update(user);
            //删除 delete
            userdao.deleteById(1);

        }catch(NamingException e){
```

```
            e.printStackTrace();
        }
    }
}
```

4. 将 ejb-project2 工程部署到 JBoss 服务器,并运行程序 JPQLUserDAOClient.java,查看运行结果。

5. 多次运行程序 JPQLUserDAOClient.java,观察运行结果是否出错,并思考其原因。

四、实验要求

1. 填写并上交实验报告,报告中应包括如下内容:
(1) 实验基本思路。
(2) 实验关键代码及相应的运行结果及截图,或相应的报错信息。
(3) 碰到的问题及解决方案或对问题的思考。
(4) 实验收获及总结。
2. 上交程序源代码,代码中应有相关注释。

扩展实验——实体关系映射操作

一、实验目的

1. 掌握多对一和一对多的实体关系映射。
2. 以多对一和一对多映射为基础,掌握一对一和多对多的实体关系映射。

二、基本知识与原理

1. 实体关系是指实体与实体之间的关系,从方向上分为单向关联和双向关联;从实体数量上分一对一、一对多(多对一)和多对多。
2. 实体关系的方向性包括单向关联和双向关联。
3. 多对一和一对多关系映射主要是指两张数据表之间存在着主键/外键关联的关系,在实体操作上存在着操作联动。
4. 多对一和一对多映射:在属性级使用 @OneToMany、@ManyToOne 和@JoinColumn 注解可定义一对多关联。
5. 一对一映射:在属性级使用 @OneToOne 和@JoinColumn 注解可定义一对一关联。
6. 多对多映射:在属性级使用 @ManyToMany 和@JoinTable 注解可定义多对多关联。注意:除少数情况需要使用到多对多关联外,一般情况如果涉及此关联,应该审视一下数据实体设计的合理性。

三、实验内容及步骤

1. 在 EIS 数据库中创建一个 department 表并修改 userlist 表,具体结构如下:

(1) 新建 department 表。

字 段 名 称	类　　型	中　文　名
departmentid	int	部门编号，主键
departmentname	varchar(50)	部门名称

```
CREATE TABLE 'department'(
  'departmentid' int(11) AUTO_INCREMENT,
  'departmentname' varchar(50) DEFAULT NULL,
  PRIMARY KEY ('departmentid')
) ENGINE = InnoDB DEFAULT CHARSET = gb2312;
```

(2) 在 userlist 表中增加一个字段 departmentid 及其外键关联。

字 段 名 称	类　　型	中　文　名
departmentid	int	部门编号，外键（与 department 表的 departmentid 关联）

```
alter table userlist add departmentid int;
alter table userlist add constraint FK_USER_DEPARTMENT foreign key (departmentid) references department(departmentid);
```

2. 修改 ejb-project2 工程：

(1) 修改 cn.edu.zjut.ejb 包下名称为 User.java 的实体 Bean，具体代码如下：

```java
package cn.edu.zjut.ejb;

import java.io.Serializable;
import javax.persistence.CascadeType;
import javax.persistence.Column;
import javax.persistence.Entity;
import javax.persistence.GeneratedValue;
import javax.persistence.GenerationType;
import javax.persistence.Id;
import javax.persistence.JoinColumn;
import javax.persistence.ManyToOne;
import javax.persistence.Table;

@SuppressWarnings("serial")
@Entity
@Table(name = "userlist")

public class User implements Serializable {

    @Id
    @GeneratedValue(strategy = GenerationType.AUTO)
    private Integer userid;

    @Column(name = "username", length = 50)
    private String username;
```

```java
    @Column(name = "userpwd", length = 50)
    private String userpwd;

    @Column(name = "age")
    private Integer age;

    @JoinColumn(name = "departmentid")
    private Department department;

    public Integer getUserid() {
        return userid;
    }
    public void setUserid(Integer userid) {
        this.userid = userid;
    }

    public String getUsername() {
        return username;
    }
    public void setUsername(String username) {
        this.username = username;
    }

    public String getUserpwd() {
        return userpwd;
    }
    public void setUserpwd(String userpwd) {
        this.userpwd = userpwd;
    }

    public Integer getAge() {
        return age;
    }

    public void setAge(Integer age) {
        this.age = age;
    }

    /*此类型与Department为"多对一关联",通过@ManyToOne注解指名.通过departmentid字段(外键)与Department相关联.*/

    @ManyToOne(cascade = CascadeType.ALL, optional = false)
    public Department getDepartment() {
        return department;
    }
    public void setDepartment(Department department) {
        this.department = department;
    }
}
```

(2) 在cn.edu.zjut.ejb包下新建名称为Department.java的实体Bean,具体代码如下:

```java
package cn.edu.zjut.ejb;

import java.io.Serializable;
import java.util.Date;
import java.util.HashSet;
import java.util.Set;

import javax.persistence.CascadeType;
import javax.persistence.Column;
import javax.persistence.Entity;
import javax.persistence.FetchType;
import javax.persistence.GeneratedValue;
import javax.persistence.GenerationType;
import javax.persistence.Id;
import javax.persistence.OneToMany;
import javax.persistence.OrderBy;
import javax.persistence.Table;
import javax.persistence.Temporal;
import javax.persistence.TemporalType;

@SuppressWarnings("serial")
@Entity
@Table(name = "department")

public class Department implements Serializable {
    private Integer departmentid;
    private String departmentname;
    private Set<User> users = new HashSet<User>();

    public Department() { }

    @Id
    @GeneratedValue(strategy = GenerationType.AUTO)
    public Integer getDepartmentid() {
        return departmentid;
    }
    public void setDepartmentid(Integer departmentid) {
        this.departmentid = departmentid;
    }

    @Column(name = "departmentname")
    public String getDepartmentname() {
        return departmentname;
    }
    public void setDepartmentname(String departmentname) {
        this.departmentname = departmentname;
    }

    /* Department 和 User 为一对多关系，而且当前类型对应的是 User 的 Department 数据成员且实
    现所有的级联操作，并且加载集合中的数据(子项)为延迟加载 */
    @OneToMany(mappedBy = "department", cascade = CascadeType.ALL, fetch = FetchType.LAZY)
```

实验十二 实体Bean——用实体Bean实现用户信息的持久化

```java
@OrderBy(value = "userid ASC")//加入到集合中的 User 对象按照 userid 排序
    public Set<User> getUsers() {
    return users;
    }
    public void setUsers(Set<User> users) {
    this.users = users;
    }

    //增加两个方法,作用为向 users 集合中添加和删除 User 对象
    public void addNewUser(User user){
    if(!this.users.contains(user)){
      this.users.add(user);
      user.setDepartment(this);
      }
    }
    public void removeUser(User user){
      user.setDepartment(null);
      this.users.remove(user);
      }
}
```

(3) 在 cn.edu.zjut.ejb.dao 包下新建 OneToManyDAORemote.java 接口,具体代码如下:

```java
package cn.edu.zjut.ejb.dao;

import java.util.List;
import javax.ejb.Remote;
import cn.edu.zjut.ejb.Department;
import cn.edu.zjut.ejb.User;

@Remote
public interface OneToManyDAORemote {
    public boolean inserDepartment(Department department);
    public Department getDepartmentById(Integer departmentid);
}
```

(4) 在 cn.edu.zjut.ejb.dao 包下新建 OneToManyDAO.java 类,具体代码如下:

```java
package cn.edu.zjut.ejb.dao;

import java.util.List;

import javax.ejb.Stateless;
import javax.persistence.*;

import cn.edu.zjut.ejb.*;

@Stateless
public class OneToManyDAO implements OneToManyDAORemote {
```

```java
@PersistenceContext(type = PersistenceContextType.EXTENDED, unitName = "departmentpu")
EntityManager em;

public boolean inserDepartment(Department department){
    try{
        em.persist(department);
    }catch(Exception e){
        e.printStackTrace();
        return false;
    }
    return true;
}
public Department getDepartmentById(Integer departmentid) {
    Department department = em.find(Department.class, departmentid);
    department.getUsers().size();
    return department;
}
```

（5）右击 ejb-project2 工程，选择菜单 Export→EJB Jar File 将工程部署到 JBoss 服务器。

3. 修改 ejb-project2-client 客户端工程：

（1）将工程 ejb-project2 的 User.java 和 Department.java 复制（覆盖）到 src 目录下新的 cn.edu.zjut.ejb 包中。

（2）将工程 ejb-project2 的 OneToManyDAORemote.java 复制到 src 目录下新的 cn.edu.zjut.ejb.dao 包中。

（3）在 src 目录下新建一个名称为 cn.edu.zjut.ejb.client 的包，新建一个文件 OneToManyTest.java，具体内容如下：

```java
package cn.edu.zjut.ejb.client;

import java.util.Hashtable;
import java.util.List;

import javax.naming.Context;
import javax.naming.InitialContext;
import javax.naming.NamingException;

import cn.edu.zjut.ejb.User;
import cn.edu.zjut.ejb.Department;
import cn.edu.zjut.ejb.dao.OneToManyDAORemote;

public class OneToManyTest {

private static OneToManyDAORemote lookupRemoteStatelessEjbBean() throws NamingException {

        final Hashtable jndiProperties = new Hashtable();
        jndiProperties.put(Context.URL_PKG_PREFIXES, "org.jboss.ejb.client.naming");
```

```java
        final Context context = new InitialContext(jndiProperties);
        final String appName = "";

        final String moduleName = "ejb-project2";

        final String distinctName = "";

        final String beanName = "OneToManyDAO";

        final String viewClassName = OneToManyDAORemote.class.getName();

        final String namespace = "ejb:" + appName + "/" + moduleName
            + "/" + distinctName + "/" + beanName + "!" + viewClassName;

        System.out.println(namespace);

        return (OneToManyDAORemote) context.lookup(namespace);

    }
    public static void main(String[] args) {
        //TODO Auto-generated method stub
        try{
            OneToManyDAORemote o2mdao = lookupRemoteStatelessEjbBean();
            Department department = new Department();
            department.setDepartmentname("行政部");
            User u1 = new User();
            u1.setUsername("张小平");
            User u2 = new User();
            u2.setUsername("张小青");
            department.addNewUser(u1);
            department.addNewUser(u2);
            o2mdao.inserDepartment(department);
        }catch(NamingException e){
            e.printStackTrace();
        }
    }
}
```

4. 启动 JBoss7.1.1Final 服务器,运行 OneToManyTest.java 文件,并查看数据库 EIS 中表 userlist 和 department 的数据列表变化,加深对实体映射操作的理解。

说明:如果在执行 OneToManyTest.java 的过程中出现类似如下异常信息:

```
Exception in thread "main"
java.lang.reflect.UndeclaredThrowableException
    at com.sun.proxy.$Proxy0.inserDepartment(Unknown Source)
    at cn.edu.zjut.ejb.client.OneToManyTest.main(OneToManyTest.java:52)
Caused by: java.lang.ClassNotFoundException: com.mysql.jdbc.MysqlDataTruncation
    at java.net.URLClassLoader$1.run(Unknown Source)
    at java.net.URLClassLoader$1.run(Unknown Source)
    at java.security.AccessController.doPrivileged(Native Method)
    at java.net.URLClassLoader.findClass(Unknown Source)
```

请打开 EIS 数据库,将 userlist 表中的 department 字段数据类型(hibernate 新增)修改为 MEDIUMBLOB,再重新运行 OneToManyTest.java 文件。

四、实验要求

1. 填写并上交实验报告,报告中应包括如下内容:
（1）运行结果截图;
（2）修改后的关键代码、相应的运行结果或报错信息;
（3）根据实验过程,总结实体映射的开发过程;
（4）碰到的问题及思考;
（5）实验收获及总结。

2. 上交程序源代码,代码中应有相关注释。

3. 尝试参照上述实验进行一对一和多对多的实体映射开发。

实验十三

消息驱动 Bean——登录用户支付消息的分发应用

基础实验——处理点对点消息

一、实验目的

1. 掌握消息驱动 Bean 的工作原理。
2. 掌握消息驱动 Bean 的开发。
3. 掌握在 JBoss 7.1 服务器上部署消息驱动 Bean。
4. 掌握点对点消息的处理。

二、基本知识与原理

1. JMS 消息服务(Java Message Service)应用程序接口能用在两个应用程序之间进行异步通信。

2. JMS 消息服务由 JMS 消息服务器(Server)、JMS 生产者(Producer)、JMS 消费者(Consumer)、JMS 消息(Message-Driven Bean)和 JMS 队列(Queue)等元素构成。

3. JMS 消息有两种类型：队列(Queue)和主题(Topic)。

4. 点对点消息(Point To Point)指一个生产者向一个特定的队列发布消息，一个消费者从该队列中读取消息。生产者不需要在接收者消费该消息期间处于运行状态，消费者也同样不需要在消息发送时处于运行状态。

5. 消息驱动 Bean(Message-Driven Bean,MDB)是设计用来专门处理基于消息请求的组件，一个 MDB 类必须实现 MessageListener 接口。消息驱动 bean 是一个异步消息使用者。当 JMS 消息到达时，容器激发消息驱动 bean，并调用 MDB 的 onMessage()方法来处理该消息。

三、实验内容及步骤

1. 新建一个名称为 ejb-project3 的 EJB 项目，并在此项目下新建一个名称为 cn.edu.zjut.ejb 的包。

2. 在 cn.edu.zjut.ejb 包下新建一个名称为 MQ.java 的消息驱动 Bean：

(1) 右击 cn.edu.zjut.ejb 包，选择 new→Message-Driven Bean 选项，如图 13-1 所示。

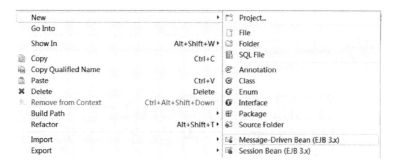

图 13-1　新建消息驱动 Bean

（2）输入 Class Name 为 MQ，选择 Destination Type 为 Queue，单击 Finish 按钮完成创建。

（3）编写 MQ.java 具体代码如下：

```java
package cn.edu.zjut.ejb;

import javax.ejb.ActivationConfigProperty;
import javax.ejb.MessageDriven;
import javax.jms.Message;
import javax.jms.MessageListener;
import javax.jms.TextMessage;

@MessageDriven(
        activationConfig = {
                @ActivationConfigProperty(
                propertyName = "destinationType", propertyValue = "javax.jms.Queue"),
                @ActivationConfigProperty(
                propertyName = "destination", propertyValue = "/queue/myqueue"),
                 @ActivationConfigProperty(propertyName = "acknowledgeMode", propertyValue = "Auto-acknowledge")
        })
public class MQ implements MessageListener {
    public MQ() { }

    public void onMessage(Message message) {
        try{
            TextMessage tmsg = (TextMessage)message;
            System.out.println("收到消息：" + tmsg.getText());
        }catch(Exception e){
            e.printStackTrace();
        }
    }
}
```

3. 更改 JBoss7.1.1.Final 服务器的启动模式，打开 JBoss 主目录下的 standalone\deployments 子目录，将 standalone.xml 重命名为 standalone-bak.xml，再将 standalone-full.xml 命名为 standalone.xml（说明：JBoss7.1.1.Final 只有在 standalone-full 配置下才

实验十三 消息驱动 Bean——登录用户支付消息的分发应用

会启动 JMS 消息服务）。

4. 打开 standalone.xml 文件，找到 hornetq-server 节点的子节点 jms-destinations，新增消息队列 myQueue 配置，具体如下：

```
<jms-destinations>
            <jms-queue name="myQueue">
                <entry name="queue/myqueue"/>
                <entry name="java:jboss/exported/jms/queue/myqueue"/>
            <jms-queue name="testQueue">
                <entry name="queue/test"/>
                <entry name="java:jboss/exported/jms/queue/test"/>
            </jms-queue>

            </jms-queue>
            <jms-topic name="testTopic">
                <entry name="topic/test"/>
                <entry name="java:jboss/exported/jms/topic/test"/>
            </jms-topic>
</jms-destinations>
```

5. 重新启动 JBoss 服务器。

6. 右击 ejb-project3 工程，在弹出的菜单中选择 Export→EJB JAR file，在弹出的窗口中单击 Browse 按钮，并在目录浏览窗口中选择 JBoss 主目录下的 standalone\deployments 子目录为部署地址 Destination（例如 C:\jboss-as-7.1.1.Final\standalone\deployments\ejb-project3.jar），当 console 窗口输出如下类似信息，表示消息驱动 Bean 已经部署成功：

```
INFO [org.jboss.as.server.deployment] (MSC service thread 1-4) JBAS015876: Starting deployment of "ejb-project3.jar"
INFO [org.jboss.as.ejb3] (MSC service thread 1-3) JBAS014142: Started message driven bean 'MQ' with 'hornetq-ra' resource adapter
INFO [org.jboss.as.server] (DeploymentScanner-threads - 1) JBAS018559: Deployed "ejb-project3.jar"
```

7. 创建 EJB 客户端工程。

（1）新建一个名称为 ejb-project3-client 的 Java Project 工程，在 ejb-project3-client 的 Java Build Path→Libraries 中添加 External JAR 库文件 jboss-client.jar，该文件所在位置为 %JBOSS_HOME%\bin\client。

（2）在 src 目录下新建一个名称为 cn.edu.zjut.ejb.client 的包，新建一个文件 MQTest.java，具体内容如下：

```
package cn.edu.zjut.ejb.client;

import java.util.Properties;

import javax.jms.Connection;
import javax.jms.ConnectionFactory;
import javax.jms.Destination;
import javax.jms.MessageProducer;
```

```java
import javax.jms.Session;
import javax.jms.TextMessage;
import javax.naming.Context;
import javax.naming.InitialContext;

public class MQTest {

    private static final String DEFAULT_MESSAGE = "这是第一条消息";
    private static final String DEFAULT_CONNECTION_FACTORY = "jms/RemoteConnectionFactory";
    private static final String DEFAULT_DESTINATION = "jms/queue/myqueue";

    private static final String DEFAULT_USERNAME = "appuser";
    private static final String DEFAULT_PASSWORD = "apppwd";
    private static final String INITIAL_CONTEXT_FACTORY = "org.jboss.naming.remote.client.InitialContextFactory";
    private static final String PROVIDER_URL = "remote://127.0.0.1:4447";

    public static void main(String[] args){
        Context context = null;
        Connection connection = null;
        try {
            //设置上下文的 JNDI 查找
            final Properties env = new Properties();
            env.put(Context.INITIAL_CONTEXT_FACTORY, INITIAL_CONTEXT_FACTORY);
                                                                      //JNDI 驱动的类名
            //Context 服务提供者的 URL 及命名服务提供者的 URL
            env.put(Context.PROVIDER_URL, PROVIDER_URL);
            //应用用户的登录名、密码
            env.put(Context.SECURITY_PRINCIPAL, DEFAULT_USERNAME);
            env.put(Context.SECURITY_CREDENTIALS, DEFAULT_PASSWORD);
            //获取 InitialContext 对象
            context = new InitialContext(env);
            ConnectionFactory connectionFactory = (ConnectionFactory) context.lookup(DEFAULT_CONNECTION_FACTORY);
            Destination destination = (Destination) context.lookup(DEFAULT_DESTINATION);

            //创建 JMS 连接、会话、生产者
            connection = connectionFactory.createConnection(DEFAULT_USERNAME, DEFAULT_PASSWORD);
            Session session = connection.createSession(false, Session.AUTO_ACKNOWLEDGE);
            MessageProducer producer = session.createProducer(destination);
            connection.start();
            //发送消息
            TextMessage message = null;
            message = session.createTextMessage(DEFAULT_MESSAGE);
            producer.send(message);
            System.out.println("消息已发送");
        } catch (Exception e) {
            e.printStackTrace();
        }
    }
}
```

8. 运行 ejb-project3-client 工程下的 MQTest.java 文件,如果 console 输出如下类似信息,则表示 JBoss 服务器已经接收到了客户端发送的消息,具体如下:

```
INFO [stdout] (Thread-35 (HornetQ-client-global-threads-6751374)) 收到消息:这是第一条消息
INFO [org.jboss.as.naming] (Remoting "jack-pc" task-3) JBAS011806: Channel end notification received, closing channel Channel ID 077ec9cd (inbound) of Remoting connection 01b95e97 to null
```

四、实验要求

1. 填写并上交实验报告,报告中应包括如下内容:
(1) 运行结果截图;
(2) 修改后的关键代码、相应的运行结果或报错信息;
(3) 根据实验过程,总结消息队列的配置和消息驱动 Bean 的开发;
(4) 碰到的问题及思考;
(5) 实验收获及总结。
2. 上交程序源代码,代码中应有相关注释。

提高实验——处理发布/订阅消息

一、实验目的

掌握发布/订阅消息的处理。

二、基本知识与原理

1. 发布/订阅模式(Publish/subscribe 或 pub/sub)是一种消息模式,消息的发送者(发布者)不是计划发送消息给特定接收者(订阅者),而是将发布的消息分为不同的主题(Topic),而无须知道这些主题有什么样的订阅者。

2. 发布者和订阅者之间存在时间依赖性。订阅者必须保持持续的活动状态以接收消息,除非订阅者建立了持久的订阅。

三、实验内容及步骤

1. 修改 ejb-project3 工程:
(1) 右击 cn.edu.zjut.ejb 包,选择 New→Message-Driven Bean 选项,在弹出的窗口中输入 Class Name 为 MT,选择 Destination Type 为 Topic,然后单击 Finish 按钮完成 MT.java 的创建。
(2) 编写 MT.java 具体代码如下:

```
package cn.edu.zjut.ejb;

import javax.ejb.ActivationConfigProperty;
import javax.ejb.MessageDriven;
import javax.jms.Message;
```

```java
import javax.jms.MessageListener;
import javax.jms.TextMessage;

@MessageDriven(
        activationConfig = {
                @ActivationConfigProperty(
                propertyName = "destinationType", propertyValue = "javax.jms.Topic"),
                @ActivationConfigProperty(
                propertyName = "destination", propertyValue = "/topic/myTopic")
        })
public class MT implements MessageListener {
    public MT() { }

    public void onMessage(Message message) {
    try{
        TextMessage tmsg = (TextMessage)message;
        System.out.println("收到主题:" + tmsg.getText());
    }catch(Exception e){
            e.printStackTrace();
    }
    }
}
```

(3) 打开 standalone.xml 文件，找到 hornetq-server 节点的子节点 jms-destinations，新增消息主题 myTopic 配置，具体如下：

```xml
<jms-destinations>
                    <jms-queue name="testQueue">
                        <entry name="queue/test"/>
                        <entry name="java:jboss/exported/jms/queue/test"/>
                    </jms-queue>
                    <jms-queue name="myQueue">
                        <entry name="queue/myqueue"/>
                        <entry name="java:jboss/exported/jms/queue/myqueue"/>
                    </jms-queue>
                    <jms-topic name="testTopic">
                        <entry name="topic/test"/>
                        <entry name="java:jboss/exported/jms/topic/test"/>
                    </jms-topic>
                    <jms-topic name="myTopic">
                        <entry name="topic/mytopic"/>
                        <entry name="java:jboss/exported/jms/topic/mytopic"/>
                    </jms-topic>
</jms-destinations>
```

(4) 重新启动 JBoss7.1.1.Final 服务器。

(5) 右击 ejb-project3 工程，在弹出的菜单中选择 Export→EJB JAR file，重新部署工程。

2. 修改 ejb-project3-client 工程。在 cn.edu.zjut.ejb.client 包新建一个文件 MTTest.java，具体内容如下：

```java
package cn.edu.zjut.ejb.client;

import java.util.Properties;

import javax.jms.TopicConnection;
import javax.jms.TopicConnectionFactory;
import javax.jms.TopicPublisher;
import javax.jms.Session;
import javax.jms.Topic;
import javax.jms.TopicSession;
import javax.jms.TextMessage;
import javax.naming.Context;
import javax.naming.InitialContext;

public class MTTest {

    private static final String DEFAULT_MESSAGE = "这是第一个消息主题";
    private static final String DEFAULT_CONNECTION_FACTORY = "jms/RemoteConnectionFactory";
    private static final String DEFAULT_DESTINATION = "jms/topic/mytopic";

    private static final String DEFAULT_USERNAME = "appuser";
    private static final String DEFAULT_PASSWORD = "apppwd";
    private static final String INITIAL_CONTEXT_FACTORY = "org.jboss.naming.remote.client.InitialContextFactory";
    private static final String PROVIDER_URL = "remote://127.0.0.1:4447";

    public static void main(String[] args){
        Context context = null;
        TopicConnection connection = null;
        try {
            //设置上下文的 JNDI 查找
            final Properties env = new Properties();
            env.put(Context.INITIAL_CONTEXT_FACTORY, INITIAL_CONTEXT_FACTORY);
                                                                            //JNDI 驱动的类名
            //Context 服务提供者的 URL 及命名服务提供者的 URL
            env.put(Context.PROVIDER_URL, PROVIDER_URL);
            //应用用户的登录名、密码
            env.put(Context.SECURITY_PRINCIPAL, DEFAULT_USERNAME);
            env.put(Context.SECURITY_CREDENTIALS, DEFAULT_PASSWORD);
            //获取 InitialContext 对象
            context = new InitialContext(env);
            TopicConnectionFactory connectionFactory = (TopicConnectionFactory) context.lookup(DEFAULT_CONNECTION_FACTORY);
            Topic destination = (Topic) context.lookup(DEFAULT_DESTINATION);
            //创建 JMS 连接、会话、发布者
            connection = connectionFactory.createTopicConnection(DEFAULT_USERNAME, DEFAULT_PASSWORD);
            TopicSession session = connection.createTopicSession(false, Session.AUTO_ACKNOWLEDGE);
            TopicPublisher publisher = session.createPublisher(destination);
            connection.start();
```

```
            //发送消息
            TextMessage message = null;
            message = session.createTextMessage(DEFAULT_MESSAGE);
            publisher.publish(message);
            System.out.println("消息主题已发送");
        } catch (Exception e) {
            e.printStackTrace();
        }
    }
}
```

3. 运行 ejb-project3-client 工程下的 MTTest.java 文件,如果 console 输出如下类似信息,则表示 JBoss 服务器已经接收到了客户端发送的消息,具体如下:

```
INFO [stdout] (Thread-1 (HornetQ-client-global-threads-2027939))
收到主题:这是第一个消息主题
```

四、实验要求

1. 填写并上交实验报告,报告中应包括:
(1) 运行结果截图;
(2) 修改后的关键代码、相应的运行结果或报错信息;
(3) 根据实验过程,总结主题的配置及其消息驱动 Bean 的开发;
(4) 碰到的问题及思考;
(5) 实验收获及总结。
2. 上交程序源代码,代码中应有相关注释。

扩展实验——支付消息的同步和异步订阅

一、实验目的

1. 掌握发布和订阅模式的开发与配置。
2. 掌握同步和异步订阅的区别。

二、基本知识与原理

发布/订阅模式支持向一个特定的消息主题发布消息。多个订阅者可接收来自特定消息主题的消息。发布者和订阅者之间存在时间依赖性。发布者需要建立一个订阅(subscription),以便客户能够订阅。订阅者必须保持持续的活动状态以接收消息。

三、实验内容及步骤

1. 修改 ejb-project3 工程。右击 cn.edu.zjut.ejb 包,选择 new→Message-Driven Bean 选项,在弹出的窗口中输入 Class Name 为 UserPayment,选择 Destination Type 为 Topic,然后单击 Finish 按钮完成创建。
2. 编写发布用户支付消息的 UserPayment.java 具体代码如下:

实验十三 消息驱动 Bean —— 登录用户支付消息的分发应用

```java
package cn.edu.zjut.ejb.client;

import java.util.Properties;

import javax.jms.TopicConnection;
import javax.jms.TopicConnectionFactory;
import javax.jms.TopicPublisher;
import javax.jms.Session;
import javax.jms.Topic;
import javax.jms.TopicSession;
import javax.jms.TextMessage;
import javax.naming.Context;
import javax.naming.InitialContext;

public class UserPayment {

    private static final String DEFAULT_MESSAGE = "用户支付消息";
    private static final String DEFAULT_CONNECTION_FACTORY = "jms/RemoteConnectionFactory";
    private static final String DEFAULT_DESTINATION = "jms/topic/mytopic";

    private static final String DEFAULT_USERNAME = "appuser";
    private static final String DEFAULT_PASSWORD = "apppwd";
    private static final String INITIAL_CONTEXT_FACTORY = "org.jboss.naming.remote.client.InitialContextFactory";
    private static final String PROVIDER_URL = "remote://127.0.0.1:4447";

    public static void main(String[] args){
        Context context = null;
        TopicConnection connection = null;
        try {
            //设置上下文的 JNDI 查找
            final Properties env = new Properties();
            //JNDI 驱动的类名
            env.put(Context.INITIAL_CONTEXT_FACTORY, INITIAL_CONTEXT_FACTORY);
            //Context 服务提供者的 URL 及命名服务提供者的 URL
            env.put(Context.PROVIDER_URL, PROVIDER_URL);
            //应用用户的登录名、密码
            env.put(Context.SECURITY_PRINCIPAL, DEFAULT_USERNAME);
            env.put(Context.SECURITY_CREDENTIALS, DEFAULT_PASSWORD);
            //获取 InitialContext 对象
            context = new InitialContext(env);
            TopicConnectionFactory connectionFactory = (TopicConnectionFactory) context.lookup(DEFAULT_CONNECTION_FACTORY);
            Topic destination = (Topic) context.lookup(DEFAULT_DESTINATION);
            //创建 JMS 连接、会话、发布者
            connection = connectionFactory.createTopicConnection(DEFAULT_USERNAME, DEFAULT_PASSWORD);
            TopicSession session = connection.createTopicSession(false, Session.AUTO_ACKNOWLEDGE);
            **TopicPublisher publisher = session.createPublisher(destination);**
            connection.start();
```

```
            //发送消息
            int count = 0;
            while(count <= 10){
            TextMessage message = null;
            message = session.createTextMessage(DEFAULT_MESSAGE + count);
                publisher.publish(message);
                System.out.println(DEFAULT_MESSAGE + count + ">>已发送");
                Thread.sleep(2000);
                count++;
            }
        }catch (Exception e) {
        e.printStackTrace();
        }
    }
}
```

3. 类似地，编写接收消息的 WarehouseDept.java(仓储部门)具体代码如下：

```
package cn.edu.zjut.ejb.client;

import java.util.Properties;

import javax.jms.TopicConnection;
import javax.jms.TopicConnectionFactory;
import javax.jms.TopicSubscriber;
import javax.jms.Session;
import javax.jms.Topic;
import javax.jms.TopicSession;
import javax.jms.TextMessage;
import javax.naming.Context;
import javax.naming.InitialContext;

public class WarehouseDept {
    private static final String DEFAULT_CONNECTION_FACTORY = "jms/RemoteConnectionFactory";
    private static final String DEFAULT_DESTINATION = "jms/topic/mytopic";

    private static final String DEFAULT_USERNAME = "appuser";
    private static final String DEFAULT_PASSWORD = "apppwd";
    private static final String INITIAL_CONTEXT_FACTORY = "org.jboss.naming.remote.client.InitialContextFactory";
    private static final String PROVIDER_URL = "remote://127.0.0.1:4447";

    public static void main(String[] args){
        Context context = null;
        TopicConnection connection = null;
        try {
            //设置上下文的 JNDI 查找
            final Properties env = new Properties();
            //JNDI 驱动的类名
            env.put(Context.INITIAL_CONTEXT_FACTORY, INITIAL_CONTEXT_FACTORY);
            //Context 服务提供者的 URL 及命名服务提供者的 URL
```

```java
            env.put(Context.PROVIDER_URL, PROVIDER_URL);
            //应用用户的登录名、密码
            env.put(Context.SECURITY_PRINCIPAL, DEFAULT_USERNAME);
            env.put(Context.SECURITY_CREDENTIALS, DEFAULT_PASSWORD);
            //获取 InitialContext 对象
            context = new InitialContext(env);
             TopicConnectionFactory connectionFactory = (TopicConnectionFactory) context.lookup(DEFAULT_CONNECTION_FACTORY);
            Topic destination = (Topic) context.lookup(DEFAULT_DESTINATION);
            //创建 JMS 连接、会话、订阅者
            connection = connectionFactory.createTopicConnection(DEFAULT_USERNAME, DEFAULT_PASSWORD);
            TopicSession session = connection.createTopicSession(false, Session.AUTO_ACKNOWLEDGE);
            TopicSubscriber subscriber = session.createSubscriber(destination);
            connection.start();
            //接收消息
            TextMessage message = null;
            message = (TextMessage) subscriber.receive();
            System.out.println("同步订阅收到消息:" + message.getText());

        } catch (Exception e) {
            e.printStackTrace();
        }
    }
}
```

4. 同理，编写接收消息的 FinancialDept.java(财务部门)，具体代码如下：

```java
package cn.edu.zjut.ejb.client;

import java.util.Properties;

import javax.jms.TopicConnection;
import javax.jms.TopicConnectionFactory;
import javax.jms.TopicSubscriber;
import javax.jms.Session;
import javax.jms.Topic;
import javax.jms.TopicSession;
import javax.jms.Message;
import javax.jms.TextMessage;
import javax.jms.MessageListener;
import javax.naming.Context;
import javax.naming.InitialContext;

public class FinancialDept implements MessageListener{
    private static final String DEFAULT_CONNECTION_FACTORY = "jms/RemoteConnectionFactory";
    private static final String DEFAULT_DESTINATION = "jms/topic/mytopic";

    private static final String DEFAULT_USERNAME = "appuser";
    private static final String DEFAULT_PASSWORD = "apppwd";
```

```java
        private static final String INITIAL_CONTEXT_FACTORY = "org.jboss.naming.remote.client.InitialContextFactory";
        private static final String PROVIDER_URL = "remote://127.0.0.1:4447";

        private int EXPECTED_MESSAGE_COUNT = 2;
        private int messageCount = 0;

    public boolean expectMoreMessage(){
        return messageCount < EXPECTED_MESSAGE_COUNT;
    }
    public void onMessage(Message m) {
        //TODO Auto-generated method stub
        System.out.println("onMessage");
        try{
            TextMessage msg = (TextMessage) m;
            System.out.println("异步订阅收到主题:" + msg.getText());
        } catch (Exception e) {
            //TODO Auto-generated catch block
            e.printStackTrace();
        }
        messageCount++;
    }
    public static void main(String[] args){
        Context context = null;
        TopicConnection connection = null;
        try {
            //设置上下文的JNDI查找
            final Properties env = new Properties();
            //JNDI驱动的类名
            env.put(Context.INITIAL_CONTEXT_FACTORY, INITIAL_CONTEXT_FACTORY);
            //Context服务提供者的URL及命名服务提供者的URL
            env.put(Context.PROVIDER_URL, PROVIDER_URL);
            //应用用户的登录名、密码
            env.put(Context.SECURITY_PRINCIPAL, DEFAULT_USERNAME);
            env.put(Context.SECURITY_CREDENTIALS, DEFAULT_PASSWORD);
            //获取InitialContext对象
            context = new InitialContext(env);
            TopicConnectionFactory connectionFactory = (TopicConnectionFactory) context.lookup(DEFAULT_CONNECTION_FACTORY);
            Topic destination = (Topic) context.lookup(DEFAULT_DESTINATION);
            //创建JMS连接、会话、订阅者
            connection = connectionFactory.createTopicConnection(DEFAULT_USERNAME, DEFAULT_PASSWORD);
            TopicSession session = connection.createTopicSession(false, Session.AUTO_ACKNOWLEDGE);
            TopicSubscriber subscriber = session.createSubscriber(destination);

            int MAX_TRIES = 100;
            int tryCount = 0;

            FinancialDept atr = new FinancialDept();
```

```
            subscriber.setMessageListener(atr);    //注意,这里注册一个监听
            connection.start();
            //试图接收消息
            System.out.println("准备接收消息(最长时限 100 秒): ");
            while(atr.expectMoreMessage() && (tryCount<MAX_TRIES)){

                try {
                    Thread.sleep(1000);
                } catch (InterruptedException e) {
                    //TODO Auto-generated catch block
                    e.printStackTrace();
                }
                tryCount++;
            }

        } catch (Exception e) {
          e.printStackTrace();
        }
    }
}
```

5. 启动 JBoss 服务器。

6. 首先运行 WarehouseDept.java 和 FinancialDept.java,然后运行 UserPayment.java,并查看3个程序的运行结果,掌握消息传递在不同子系统中的应用原理,掌握同步和异步订阅的差异。

四、实验要求

1. 填写并上交实验报告,报告中应包括如下内容:
（1）运行结果截图；
（2）修改后的关键代码、相应的运行结果或报错信息；
（3）根据实验过程,总结发布/订阅可以服务哪些应用场景；
（4）碰到的问题及思考；
（5）实验收获及总结。
2. 上交程序源代码,代码中应有相关注释。
3. 类似地完成补充实验：建立一个用户付款的 Web 页面,用户单击"提交"按钮将付款成功的消息分别发送给销售部门和仓储配货部门。

第四部分　Java EE综合应用开发

- 实验十四　综合应用
　　——基于SSH的网上书城
- 实验十五　综合应用
　　——基于EJB的网上书城

实验十四

综合应用
——基于 SSH 的网上书城

一、实验目的

1. 掌握 Spring 框架、Hibernate 框架、Struts 框架的基本开发方法以及 3 个框架的整合方法,理解 3 个框架在系统各层次中所起的作用及其相互之间的关系。

2. 通过综合应用理解 Java EE 所表达的软件架构和设计思想。

3. 能综合运用所学的知识,应用轻量级 Java EE 架构,完成系统的分析、设计、开发、部署、调试、测试,培养解决实际问题的能力。

二、实验环境

1. Java 环境:jdk 1.6 以上版本。
2. 集成开发环境:Eclipse 或 Myeclipse。
3. Web 服务器:Tomcat 6.0 以上版本。
4. 轻量级 Java EE 框架:Struts2、Hibernate3 以上版本、Spring3 以上版本。
5. 数据库:MySQL 5.0 以上版本或 SQL Server 2000 以上版本。

三、实验用例分析

网上书城系统共有"用户"(包括匿名用户和注册用户)、"书城管理员"和"系统管理员"三大类角色,分别在基础、提高和扩展 3 个层次的实验中加以介绍。

(一)系统基础用例

基础实验内容围绕"用户"角色展开,其用例图为图 14-1。其中,"匿名用户"拥有"用户"的基本功能;而"注册用户"除了拥有"用户"的基本功能外,还具有更新注册信息、下订单等功能。

1. 用户能在系统中完成注册,也能在没有注册登录的情况下完成根据类别查看书籍信息、根据关键字查询书籍信息、将书籍添加到购物车、查看或修改购物车信息等操作。用户

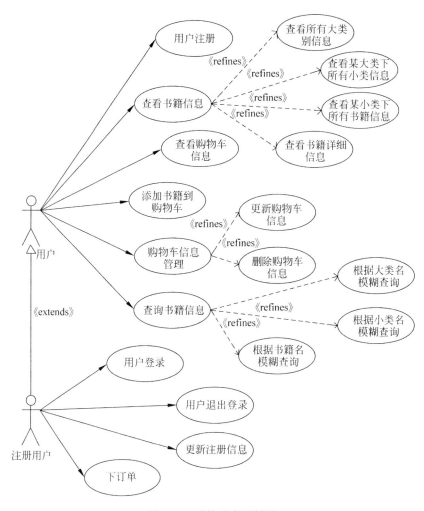

图 14-1　系统基本用例图

的详细用例描述如表 14-1～表 14-6 所示。

表 14-1　用户注册用例描述

用例标识		ID14-1	用例名称	用户注册
用例描述		用户打开网站的注册页面进行个人信息的注册,注册成功后即成为注册用户,同时,网站数据库将保存该用户的个人信息		
参与者		用户		
前置条件		用户打开网站并进入注册页面		
事件流	基本流程	① 用户打开本网站 ② 进入注册页面 ③ 在注册表单中填写用户名、密码等个人信息 ④ 单击表单的"注册"按钮,完成注册		
	扩展流程	当用户填写个人信息有误时,单击表单的"重置"按钮,清空表单信息		
	异常流程	数据库写入异常导致注册失败,返回注册页面并显示错误提示信息		
后置条件		用户注册成功并显示注册成功提示信息		

表 14-2　用户查看书籍信息用例描述

用例标识		ID14-2	用例名称	查看书籍信息
用例描述		用户打开网站页面查看书籍信息,可以查看某一大类中的所有书籍信息,可以查看大类中某一小类的所有书籍信息,也可以查看所有类型的书籍信息			
参与者		用户			
前置条件		用户打开网站主页			
事件流	基本流程	① 用户打开本网站主页 ② 单击菜单栏中的"大类别"或"小类别"等菜单项 ③ 进入相应类别页面,并显示该类别范围内的书籍信息			
	扩展流程				
	异常流程	读取数据库异常导致查看失败,返回当前页面并显示错误提示信息			
后置条件		进入相应类别页面,并显示该类别范围内的书籍信息			

表 14-3　用户查询书籍信息用例描述

用例标识		ID14-3	用例名称	查询书籍信息
用例描述		用户打开网站页面,在搜索栏中选择搜索类型并输入搜索关键词,单击"搜索"按钮后,在页面中显示搜索结果,该搜索结果应在搜索的特定类型中并包含有关键词			
参与者		用户			
前置条件		用户打开网站页面			
事件流	基本流程	① 用户打开本网站页面 ② 选择搜索栏中的搜索类型,如搜索"大类别"、"小类别"或"书籍名称"等 ③ 在搜索窗口输入搜索关键词 ④ 单击"搜索"按钮,完成搜索			
	扩展流程	若没有输入关键词,则查询所有结果			
	异常流程	读取数据库异常导致搜索失败,返回当前页面并显示错误提示信息			
后置条件		在页面中显示搜索结果,该搜索结果应在搜索的特定类型中并包含有关键词			

表 14-4　添加书籍到购物车用例描述

用例标识		ID14-4	用例名称	添加书籍到购物车
用例描述		用户在网站中选择待购的书籍,单击"添加到购物车"按钮,将待购的书籍加入到当前用户的购物车中			
参与者		用户			
前置条件		用户打开网站,进入提供书籍购买功能的页面			
事件流	基本流程	① 用户打开本网站页面 ② 进入提供书籍购买功能的页面 ③ 选择某本书籍,单击"添加到购物车"按钮,将该书籍添加到当前用户的购物车中 ④ 返回当前页面,继续选择书籍购买			
	扩展流程	若待购的书籍已存在当前用户的购物车中,则单击"添加到购物车"按钮后,购物车中该书籍的数量加 1			
	异常流程	添加购物车失败,返回当前页面并显示错误提示信息			
后置条件		当前用户的购物车中增加选择购买的书籍信息			

表 14-5　查看购物车信息用例描述

用例标识	ID14-5	用例名称	查看购物车信息
用例描述			用户进入网站的购物车页面，查看当前用户购物车中的书籍信息(包括书籍名称、单价、数量、总价等)
参与者			用户
前置条件			用户打开网站的购物车页面
事件流	基本流程		① 用户打开本网站 ② 进入购物车页面 ③ 浏览当前用户购物车中的书籍信息
	扩展流程		若当前用户的购物车中没有任何书籍，则在本页面显示提示信息
	异常流程		查看购物车信息失败，返回当前页面并显示错误提示信息
后置条件			查看购物车信息成功

表 14-6　购物车信息管理用例描述

用例标识	ID14-6	用例名称	购物车信息管理
用例描述			用户进入网站的购物车页面，修改当前用户购物车中的待购书籍数量，或删除购物车中某一本或几本书籍(即书籍数量为0)，修改成功后返回当前页面以便用户查看更新后的购物车信息
参与者			用户
前置条件			查看购物车信息
事件流	基本流程		① 用户打开本网站 ② 进入购物车页面 ③ 修改当前用户购物车中的待购书籍数量，或删除购物车中某一本或几本书籍(即书籍数量为0) ④ 修改成功后返回当前页面，以便用户查看更新后的购物车信息
	扩展流程		若当前用户的购物车中没有任何书籍，则在本页面显示提示信息
	异常流程		修改购物车失败，返回当前页面并显示错误提示信息
后置条件			修改成功后返回当前页面

2. 注册用户是用户中的一种特殊类型，是在系统中完成注册了的一类用户。注册用户除了拥有用户的基本功能外，还具有更新注册信息、下订单等功能。注册用户的详细用例描述如表 14-7～表 14-10 所示。

表 14-7　注册用户登录用例描述

用例标识	ID14-7	用例名称	注册用户登录
用例描述			用户打开网站，在登录表单中填写用户名和密码，单击"登录"按钮完成登录。若登录成功，则在当前页面显示欢迎信息；若登录失败则跳转到错误页面并显示错误提示信息(如用户名不存在、密码错误等)
参与者			注册用户
前置条件			注册用户打开本网站
事件流	基本流程		① 注册用户打开本网站 ② 在登录表单中填写用户名和密码 ③ 单击表单的"登录"按钮，完成登录
	扩展流程		当注册用户填写用户名或密码有误时，单击表单的"重置"按钮，清空表单信息
	异常流程		登录失败，跳转到错误页面并显示错误提示信息(如用户名不存在、密码错误等)
后置条件			用户登录成功并在当前页面显示欢迎信息

表 14-8　注册用户退出登录用例描述

用例标识		ID14-8	用例名称	注册用户退出登录
用例描述		已登录的用户在网站的页面中单击"注销"按钮,退出登录		
参与者		注册用户		
前置条件		注册用户已在网站中登录		
事件流	基本流程	① 已登录的注册用户打开本网站某一页面 ② 单击页面中的"注销"按钮 ③ 已登录的用户退出登录		
	扩展流程			
	异常流程	退出登录失败,跳转到错误页面并显示错误提示信息		
后置条件		用户退出登录成功并在当前页面显示登录表单		

表 14-9　更新注册信息用例描述

用例标识		ID14-9	用例名称	更新注册信息
用例描述		已登录的用户进入网站的注册信息更新页面,在表单中,对当前用户的个人信息进行更新,单击表单的"保存"按钮,完成更新		
参与者		注册用户		
前置条件		注册用户已在网站中登录,并打开本网站的注册信息更新页面		
事件流	基本流程	① 已登录的注册用户打开本网站的注册信息更新页面 ② 在表单中,对当前用户的个人信息进行更新 ③ 单击表单的"保存"按钮,完成更新		
	扩展流程	当用户更新的注册信息有误时,单击表单的"重置"按钮,清空表单信息		
	异常流程	更新注册信息失败,跳转到当前页面并显示错误提示信息		
后置条件		用户更新注册信息成功,并在当前页面显示更新后的信息		

表 14-10　下订单用例描述

用例标识		ID14-10	用例名称	下订单
用例描述		用户进入网站的购物车页面,在确认购物车信息后,单击"下订单"按钮,完成购买。若当前用户已登录,则在完成购买后显示订单信息;若未登录,则不能完成购买,而跳转到登录/注册页面		
参与者		注册用户		
前置条件		注册用户打开本网站的购物车页面		
事件流	基本流程	① 注册用户打开本网站的购物车页面 ② 查看购物车信息 ③ 确认购物车信息后,单击"下订单"按钮,完成购买 ④ 下订单成功后,页面显示本订单信息		
	扩展流程	① 当购物车信息有误时,修改购物车信息,直到确认无误 ② 若用户尚未登录,则单击"下订单"按钮后,跳转到登录/注册页面先进行登录或注册		
	异常流程	下订单失败,跳转到当前页面并显示错误提示信息		
后置条件		下订单成功,并在页面中显示订单信息		

（二）系统提高用例

提高实验内容围绕"书城管理员"角色展开,其用例图为图 14-2。

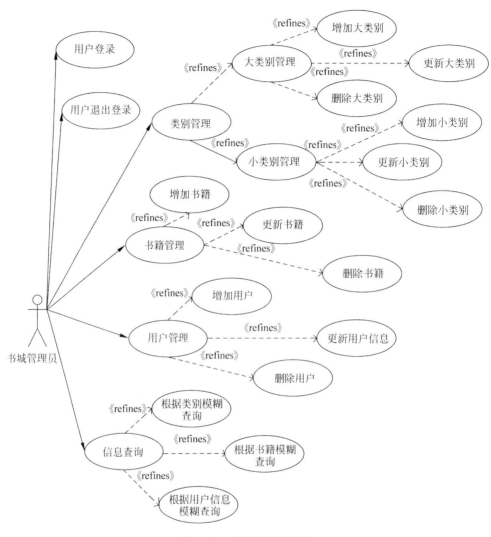

图 14-2 系统提高用例图

书城管理员在登录后,能在系统中完成类别管理、书籍管理、用户管理和信息查询等操作。书城管理员的详细用例描述如表 14-11～表 14-14 所示(用户登录和用户退出登录用例描述省略)。

表 14-11 类别管理用例描述

用例标识	ID14-11	用例名称	类别管理
用例描述	书城管理员通过登录进入网站,在菜单栏中选择"类别管理",进入类别管理页面,对书籍的类别(大类别或小类别)进行增加、删除或信息修改,操作成功后返回当前页面并显示更新后的类别信息			

续表

	参与者	书城管理员
	前置条件	书城管理员在网站登录成功
事件流	基本流程	① 书城管理员通过登录进入网站 ② 在菜单栏中选择"类别管理",进入类别管理页面 ③ 对书籍的类别(大类别或小类别)进行增加、删除或信息修改 ④ 操作成功后返回当前页面并显示更新后的类别信息
	扩展流程	
	异常流程	类别管理失败,返回当前页面并显示错误提示信息
	后置条件	操作成功,返回当前页面并显示更新后的类别信息

表 14-12　书籍管理用例描述

用例标识	ID14-12	用例名称	书籍管理
用例描述	书城管理员通过登录进入网站,在菜单栏中选择"书籍管理",进入书籍管理页面,对书籍的信息进行增加、删除或信息修改,操作成功后返回当前页面并显示更新后的书籍信息			
参与者	书城管理员			
前置条件	书城管理员在网站登录成功			
事件流	基本流程	① 书城管理员通过登录进入网站 ② 在菜单栏中选择"书籍管理",进入书籍管理页面 ③ 对书籍的信息进行增加、删除或信息修改 ④ 操作成功后返回当前页面并显示更新后的书籍信息		
	扩展流程			
	异常流程	书籍管理失败,返回当前页面并显示错误提示信息		
后置条件	操作成功,返回当前页面并显示更新后的书籍信息			

表 14-13　用户管理用例描述

用例标识	ID14-13	用例名称	用户管理
用例描述	书城管理员通过登录进入网站,在菜单栏中选择"用户管理",进入用户管理页面,对用户的信息进行增加、删除或信息修改,操作成功后返回当前页面并显示更新后的用户信息			
参与者	书城管理员			
前置条件	书城管理员在网站登录成功			
事件流	基本流程	① 书城管理员通过登录进入网站 ② 在菜单栏中选择"用户管理",进入用户管理页面 ③ 对用户的信息进行增加、删除或信息修改 ④ 操作成功后返回当前页面并显示更新后的用户信息		
	扩展流程			
	异常流程	用户管理失败,返回当前页面并显示错误提示信息		
后置条件	操作成功,返回当前页面并显示更新后的用户信息			

表 14-14　信息查询用例描述

用例标识	ID14-14	用例名称	信息查询
用例描述	书城管理员通过登录进入网站,在搜索栏中选择搜索项(如搜索类型、书籍或用户等)并输入搜索关键词,单击"搜索"按钮后,页面中显示搜索结果,该搜索结果应在特定搜索项中并包含有关键词		
参与者	书城管理员		
前置条件	书城管理员在网站登录成功		
事件流	基本流程	① 书城管理员通过登录进入网站 ② 选择搜索栏中的搜索项,如搜索类型、书籍或用户等 ③ 在搜索窗口输入搜索关键词 ④ 单击"搜索"按钮,完成搜索	
	扩展流程	若没有输入关键词,则查询所有结果	
	异常流程	读取数据库异常导致搜索失败,返回当前页面并显示错误提示信息	
后置条件	在页面中显示搜索结果,该搜索结果应在特定搜索项中并包含有关键词		

（三）系统扩展用例

扩展实验内容围绕"系统管理员"角色展开,其用例图为图 14-3。

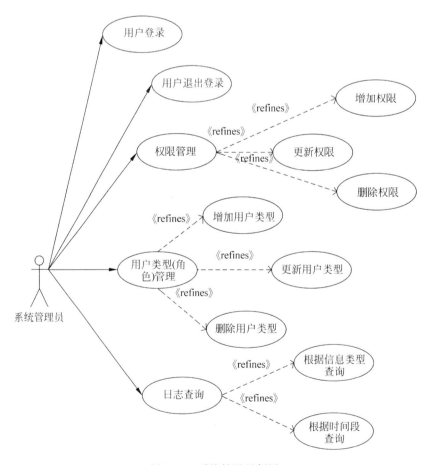

图 14-3　系统扩展用例图

系统管理员在登录后,能在系统中完成权限管理、用户类型(角色)管理和日志查询等操作。系统管理员的详细用例描述如表14-15～表14-17所示(用户登录和用户退出登录用例描述省略)。

表 14-15　权限管理用例描述

用例标识		ID14-15	用例名称	权限管理
用例描述		系统管理员通过登录进入网站,在菜单栏中选择"权限管理",进入权限管理页面,对系统权限进行增加、删除或修改。操作成功后返回当前页面并显示更新后的权限信息		
参与者		系统管理员		
前置条件		系统管理员在网站登录成功		
事件流	基本流程	① 系统管理员通过登录进入网站 ② 在菜单栏中选择"权限管理",进入权限管理页面 ③ 对系统权限进行增加、删除或修改 ④ 操作成功后返回当前页面并显示更新后的权限信息		
	扩展流程			
	异常流程	权限管理操作失败,返回当前页面并显示错误提示信息		
后置条件		操作成功,返回当前页面并显示更新后的系统权限信息		

表 14-16　用户类型(角色)管理用例描述

用例标识		ID14-16	用例名称	用户类型(角色)管理
用例描述		系统管理员通过登录进入网站,在菜单栏中选择"用户类型(角色)管理",进入用户类型(角色)管理页面,对系统角色进行增加、删除或修改某角色所拥有的权限。操作成功后返回当前页面并显示更新后的用户类型(角色)信息		
参与者		系统管理员		
前置条件		系统管理员在网站登录成功		
事件流	基本流程	① 系统管理员通过登录进入网站 ② 在菜单栏中选择"用户类型(角色)管理",进入用户类型(角色)管理页面 ③ 对系统角色进行增加、删除或修改某角色所拥有的权限 ④ 操作成功后返回当前页面并显示更新后的用户类型(角色)信息		
	扩展流程			
	异常流程	用户类型(角色)管理操作失败,返回当前页面并显示错误提示信息		
后置条件		操作成功,返回当前页面并显示更新后的用户类型(角色)信息		

表 14-17　日志查询用例描述

用例标识	ID14-17	用例名称	日志查询
用例描述	系统管理员通过登录进入网站,在菜单栏中选择"日志查询",进入日志查询页面,在搜索栏中选择搜索条件(如信息类型、时间段等)并输入搜索关键词。单击"搜索"按钮后,页面中显示搜索结果。该搜索结果应在特定搜索条件中并包含有关键词		
参与者	系统管理员		
前置条件	系统管理员在网站登录成功		

续表

事件流	基本流程	① 系统管理员通过登录进入网站 ② 在菜单栏中选择"日志查询",进入日志查询页面 ③ 选择搜索栏中的搜索条件,如信息类型、时间段等 ④ 在搜索窗口输入搜索关键词 ⑤ 单击"搜索"按钮,完成搜索
	扩展流程	若没有输入关键词,则查询所有结果
	异常流程	读取数据库异常导致搜索失败,返回当前页面并显示错误提示信息
后置条件		在页面中显示搜索结果,该搜索结果应在特定搜索条件中并包含有关键词

四、实验要求

1. 根据实验用例,以及各实验小组对系统需求的细化,完成系统设计,包括系统用例细化、系统功能模块设计、系统主要业务流程说明、系统数据库表格与 E-R 图、系统框架设计等。

2. 根据系统设计,使用 SSH 框架完成系统开发,实现系统功能。

3. 对系统功能进行简单测试,根据测试结果完善系统。

4. 撰写综合实验报告,报告中应包括系统需求分析、系统设计、系统实现和实验总结 4 大部分。

5. 上交程序源代码,代码中应有相关注释。

实验十五

综合应用
——基于 EJB 的网上书城

一、实验目的

1. 掌握 EJB 中会话 Bean、实体 Bean 和消息驱动 Bean 的基本开发方法及其相互整合的方法,理解 EJB 在系统各层次中所起的作用及其相互之间的关系。
2. 通过综合应用理解 Java EE 所表达的软件架构和设计思想。
3. 能综合运用所学的知识,应用企业级 Java EE 架构,完成系统的分析、设计、开发、部署、调试、测试,培养解决实际问题的能力。

二、实验环境

1. Java 环境:jdk 1.6 以上版本。
2. 集成开发环境:Eclipse 或 Myeclipse。
3. Web 服务器:JBoss 6.0 以上版本。
4. 企业级 Java EE 框架:EJB3.0 以上版本。
5. 数据库:MySQL 5.0 以上版本或 SQL Server 2000 以上版本。

三、实验用例分析

网上书城包括三大子系统:书籍选购子系统、书籍管理子系统和订单调度中心子系统。书籍选购子系统中的主要角色是"用户"(包括匿名用户和注册用户),"用户"可以完成对书籍的查看、加入购物车、下订单等操作。书籍管理子系统中的主要角色是"书城管理员","书城管理员"可以对书籍信息、用户信息等进行增、删、改、查的操作管理。订单调度中心子系统中的主要角色是"订单管理员","订单管理员"可以对客户提交的订单进行处理,通过 JMS 发订单给供应商,然后通过 Java Mail 来通知客户订单处理情况。对三大子系统及相应的三大类角色的介绍,分别被划分在基础、提高和扩展 3 个层次的实验内容中。

(一)系统基础用例

基础实验内容围绕书籍选购子系统和"用户"角色展开,其用例图为图 14-1。其中,"匿名用户"拥有"用户"的基本功能；而"注册用户"除了拥有"用户"的基本功能外，还具有更新注册信息、下订单等功能。"用户"的详细用例描述如表 14-1～表 14-10 所示（详见实验十四）。

（二）系统提高用例

提高实验内容围绕书籍管理子系统和"书城管理员"角色展开，其用例图为图 14-2。"书城管理员"在登录后，能在系统中完成类别管理、书籍管理、用户管理和信息查询等操作。书城管理员的详细用例描述如表 14-11～表 14-14 所示（详见实验十四）。

（三）系统扩展用例

扩展实验内容围绕订单调度中心子系统和"订单管理员"角色展开，其用例图为图 15-1。

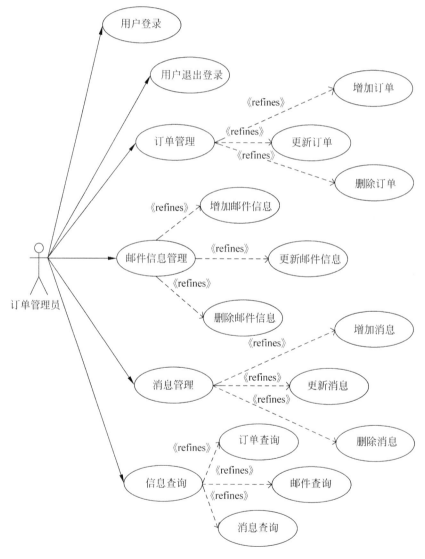

图 15-1　系统扩展用例图

订单调度中心子系统的主要功能是对客户提交的订单进行处理,通过 JMS 发订单给供应商,然后通过 Java Mail 来通知客户订单处理情况。订单管理员在登录后,能在系统中完成订单管理、邮件信息管理、消息管理和信息查询等操作。订单管理员的详细用例描述如表 15-1~表 15-4 所示(用户登录和用户退出登录用例描述省略)。

表 15-1 订单管理用例描述

用例标识		ID15-1	用例名称	订单管理
用例描述		订单管理员通过登录进入网站,在菜单栏中选择"订单管理",进入订单管理页面,对订单进行增加、删除或修改,操作成功后返回当前页面并显示更新后的订单信息。若订单状态被修改,则通过 JMS 或 Java Mail 通知供应商和客户		
参与者		订单管理员		
前置条件		订单管理员在网站登录成功		
事件流	基本流程	① 订单管理员通过登录进入网站 ② 在菜单栏中选择"订单管理",进入订单管理页面 ③ 对订单进行增加、删除或修改,特别是对订单状态(如订单确认、订单取消)的修改 ④ 操作成功后返回当前页面并显示更新后的订单信息		
	扩展流程	若订单状态被修改,则通过 JMS 或 Java Mail 通知供应商和客户		
	异常流程	订单管理操作失败,返回当前页面并显示错误提示信息		
后置条件		操作成功,返回当前页面并显示更新后的订单信息		

表 15-2 邮件信息管理用例描述

用例标识		ID15-2	用例名称	邮件信息管理
用例描述		订单管理员通过登录进入网站,在菜单栏中选择"邮件信息管理",进入邮件信息管理页面,对邮件类型、邮件内容等进行增加、删除或修改,操作成功后返回当前页面并显示更新后的邮件信息		
参与者		订单管理员		
前置条件		订单管理员在网站登录成功		
事件流	基本流程	① 订单管理员通过登录进入网站 ② 在菜单栏中选择"邮件信息管理",进入邮件信息管理页面 ③ 对邮件类型、邮件内容等进行增加、删除或修改 ④ 操作成功后返回当前页面并显示更新后的邮件信息		
	扩展流程			
	异常流程	邮件信息管理操作失败,返回当前页面并显示错误提示信息		
后置条件		操作成功,返回当前页面并显示更新后的邮件信息		

表 15-3 消息管理用例描述

用例标识	ID15-3	用例名称	消息管理
用例描述	订单管理员通过登录进入网站,在菜单栏中选择"消息管理",进入消息管理页面,对消息类型、消息内容等进行增加、删除或修改,操作成功后返回当前页面并显示更新后的消息信息		
参与者	订单管理员		
前置条件	订单管理员在网站登录成功		

续表

事件流	基本流程	① 订单管理员通过登录进入网站 ② 在菜单栏中选择"消息管理",进入消息管理页面 ③ 对消息类型、消息内容等进行增加、删除或修改 ④ 操作成功后返回当前页面并显示更新后的消息
	扩展流程	
	异常流程	消息管理操作失败,返回当前页面并显示错误提示信息
后置条件		操作成功,返回当前页面并显示更新后的消息

表 15-4　信息查询用例描述

用例标识		ID15-4	用例名称	信息查询
用例描述		订单管理员通过登录进入网站,在菜单栏中选择"信息查询",进入信息查询页面,在搜索栏中选择搜索项(如订单查询、消息查询等)并输入搜索关键词。单击"搜索"按钮后,页面中显示搜索结果。该搜索结果应在特定搜索项中并包含有关键词		
参与者		订单管理员		
前置条件		订单管理员在网站登录成功		
事件流	基本流程	① 订单管理员通过登录进入网站 ② 在菜单栏中选择"信息查询",进入信息查询页面 ③ 选择搜索栏中的搜索项,如订单查询、消息查询等 ④ 在搜索窗口输入搜索关键词 ⑤ 单击"搜索"按钮,完成搜索		
	扩展流程	若没有输入关键词,则查询所有结果		
	异常流程	读取数据库异常导致搜索失败,返回当前页面并显示错误提示信息		
后置条件		在页面中显示搜索结果,该搜索结果应在特定搜索项中并包含有关键词		

四、实验要求

1. 根据实验用例以及各实验小组对系统需求的细化,完成系统设计,包括系统用例细化、系统功能模块设计、系统主要业务流程说明、系统数据库表格与 E-R 图、系统框架设计等。

2. 根据系统设计,使用 EJB 框架完成系统开发,实现系统功能。

3. 对系统功能进行简单测试,根据测试结果完善系统。

4. 撰写综合实验报告,报告中应包括系统需求分析、系统设计、系统实现和实验总结 4 大部分。

5. 上交程序源代码,代码中应有相关注释。